MACMILLAN MATHEMATICAL GUIDES

**Linear Algebra** D. Towers
**Abstract Algebra** C. Whitehead
**Analysis** F. Hart
**Numerical Analysis** J. Turner
**Mathematical Modelling** D. Edwards and M. Hanson
**Mathematical Methods** J. Gilbert

# Guide to Analysis

F. Mary Hart

*Department of Pure Mathematics*
*University of Sheffield*

MACMILLAN

First published 1988 by
THE MACMILLAN PRESS LTD
Houndmills, Basingstoke, Hampshire RG21 2XS
and London
Companies and representatives
throughout the world

ISBN 0–333–43788–8

A catalogue record for this book is available
from the British Library

Printed in Hong Kong

Reprinted 1990, 1991, 1992

# CONTENTS

# EDITOR'S FOREWORD

Wide concern has been expressed in tertiary education about the difficulties experienced by students during their first year of an undergraduate course containing a substantial component of mathematics. These difficulties have a number of underlying causes, including the change of emphasis from an algorithmic approach at school to a more rigorous and abstract approach in undergraduate studies, the greater expectation of independent study, and the increased pace at which material is presented. The books in this series are intended to be sensitive to these problems.

Each book is a carefully selected, short, introductory text on a key area of the first-year syllabus; the areas are complementary and largely self-contained. Throughout, the pace of development is gentle, sympathetic and carefully motivated. Clear and detailed explanations are provided, and important concepts and results are stressed.

As mathematics is a practical subject which is best learned by doing it, rather than watching or reading about someone else doing it, a particular effort has been made to include a plentiful supply of worked examples, together with appropriate exercises, ranging in difficulty from the straightforward to the challenging.

When one goes fellwalking, the most breathtaking views require some expenditure of effort in order to gain access to them: nevertheless, the peak is more likely to be reached if a gentle and interesting route is chosen. The mathematical peaks attainable in these books are every bit as exhilarating, the paths are as gentle as we could find, and the interest and expectation are maintained throughout to prevent the spirits from flagging on the journey.

*Lancaster, 1987*

David A. Towers
Consultant Editor

# PREFACE

The transition from sixth form to university or polytechnic can be traumatic. Students have to adapt to different styles of teaching with lectures replacing lessons and at the same time they are often presented with material which is unlike anything they have encountered before. The provision of suitable textbooks can go some way to alleviating the problem. As far as pure mathematics is concerned, analysis, which is central to most undergraduate courses in mathematics, is likely to provide the biggest stumbling-block. The subject itself is intrinsically difficult. The final polished version we have today owes its form to years of concentrated work by many celebrated mathematicians, and students often find it very daunting to be faced with this finished product. They seldom realise just how much effort went into its making, who did the work or why they did it. This account therefore includes some historical notes and anecdotes which help to put the subject into context and, it is hoped, enliven the text for the reader.

This volume is written specifically for students meeting analysis for the first time. It is designed so that it can be used as a textbook in conjunction with a lecture course in analysis or it can be used by students wishing to teach themselves the subject—independent of any lecture course. For this reason numerous worked examples are included with very detailed solutions to enable the reader to check that the concepts and definitions are understood. When working the exercises the reader is given the opportunity to receive a little help (or inspiration!) without being given the answer, as the answers and the hints for solutions are recorded in separate sections. In practice, one exercise successfully completed by the reader (even with the help of a few hints) can teach more than several examples read from the text.

Analysis traditionally contains numerous theorems and proofs. These theorems are the tools of the trade and we cannot just dispense with large numbers of them in order either to cut down the size of the volume or to make the learning process appear easier. Neither can we simply omit their proofs. Before using a result we must know for certain that it is true. The need for proof is something which has always been recognised to a greater or lesser extent. Historical records show that the ancient Greek mathematicians living centuries before Christ acknowledged this, and, indeed, some of our present-day proofs, like the contradiction argument, stem from their work. However, the use of the link between limits of functions and sequences allows us a few economies and reduces the number of separate proofs necessary.

For many undergraduates the greatest hurdle is the problem posed by the idea of rigour. They are unsure about how much detail needs to be included and how much may reasonably be omitted. The reader of more advanced texts can be assumed to have sufficient judgement to be able to cope when details are not carefully spelled out. A more elementary text, however, needs different treatment and it seems prudent to include all the small details. This may make the definitions and theorems appear a little more cumbersome, but it leaves the reader in no doubt as to what is being assumed. The benefits far outweigh any superficial disadvantage occurring from the longer, more detailed statements. Since repetitions can be valuable for reinforcing knowledge, some statements and proofs which closely resemble earlier ones are nevertheless recorded in detail rather than being dismissed with an assertion of the form 'the proof is similar to that of...'.

Finally, I would like to record my profound thanks for all the secretarial help I have received during the preparation of this volume. My thanks go to Anne Hall and Janet Williams and particularly to Alex Cain who has undertaken the bulk of the work.

*Sheffield, 1987* F. M. Hart

# GLOSSARY OF SYMBOLS AND NOTATION

| | |
|---|---|
| $\mathbb{N}$ | Set of all natural numbers. This set consists of the numbers $0, 1, 2, 3, \ldots$ |
| $\mathbb{Z}$ | Set of all integers |
| $\mathbb{Z}^+$ | Set of all positive integers |
| $\mathbb{Q}$ | Set of all rational numbers |
| $\mathbb{R}$ | Set of all real numbers |
| $\mathbb{R}\backslash\{0\}$ | Set of all non-zero real numbers |
| $\|x\|$ | Modulus of a real number $x$. Its value is $x$ if $x \geqslant 0$ and its value is $-x$ if $x < 0$. Thus $\|x\| = +\sqrt{x^2}$ |
| $>$ | Greater than |
| $<$ | Less than |
| $\ngtr$ | Not greater than |
| $\nless$ | Not less than |
| $\geqslant$ | Greater than or equal to |
| $\leqslant$ | Less than or equal to |
| $\{x\colon \mathrm{P}(x)\}$ | The set of all $x$ such that $\mathrm{P}(x)$ holds |
| $\{a, b, c, \ldots\}$ | The set consisting of $a, b, c, \ldots$ |
| $\to$ | Tends to |
| $\not\to$ | Does not tend to |
| $\in$ | Belongs to |
| $\notin$ | Does not belong to |
| $(a_n)_{n=1}^{\infty}$ | Infinite sequence $a_1, a_2, a_3, \ldots$ whose $n$th term is $a_n$. (This notation is used when we wish to make clear which values are assumed by the subscript $n$) |
| $(a_n)$ | Infinite sequence with terms of the form $a_n$. (This notation is used when it is clear from the context which values are assumed by $n$) |
| $(a_{n_k})_{k=1}^{\infty}$ | Subsequence $a_{n_1}$, $a_{n_2}$, $a_{n_3}$, ... of the sequence $(a_n)$ |
| $\sum$ | Sum of |

|  |  |
|---|---|
| $\sum_{n=1}^{\infty} a_n$ | This is used in two senses. It is used formally to stand for the infinite series $a_1 + a_2 + a_3 + \ldots$ It is also used for the sum of the infinite series in the case in which the series is convergent. The context makes clear which meaning is appropriate |
| $\sum_{k=1}^{n} a_k$ | The sum of the $n$ terms $a_1 + a_2 + \ldots + a_n$ |
| $\gamma$ | Euler's constant |
| $f: A \to \mathbb{R}$ | Function $f$ from $A$ into $\mathbb{R}$ i.e. $f$ has domain of definition $A$ and range a subset of $\mathbb{R}$ |
| $\log x$ | The *natural* logarithm of $x$. Throughout this volume all logarithms are to the base e |
| $(a, \infty)$ | Set of all real numbers $x$ such that $x > a$ |
| $[a, \infty)$ | Set of all real numbers $x$ such that $x \geqslant a$ |
| $(-\infty, b)$ | Set of all real numbers $x$ such that $x < b$ |
| $(-\infty, b]$ | Set of all real numbers $x$ such that $x \leqslant b$ |
| $(a, b)$ | The open interval $(a, b)$. It consists of all the real numbers $x$ such that $a < x < b$ |
| $[a, b]$ | The closed interval $[a, b]$. It consists of all the real numbers $x$ such that $a \leqslant x \leqslant b$ |
| $x \to a$ | $x$ approaches $a$ (two-sided limit) |
| $x \to a^+$ | $x$ approaches $a$ through values greater than $a$ (one-sided limit) |
| $x \to a^-$ | $x$ approaches $a$ through values less than $a$ (one-sided limit) |
| $g \circ f$ | $g$ composition $f$. This is the function such that $(g \circ f)(x) = g(f(x))$ |
| $f^{-1}$ | Inverse function of $f$ |
| $f'$ | Derivative of $f$ |
| IVT | Intermediate value theorem |
| MVT | Mean value theorem |

# 1 NUMBERS AND NUMBER SYSTEMS

This is a chapter in which we have another look at some familiar number systems and meet some rather unexpected ideas.

## 1.1 NATURAL NUMBERS

We begin with the familiar numbers $0, 1, 2, 3, 4, \ldots$, which we have used for counting since early childhood. Not surprisingly, this set of numbers is called the set of natural numbers and it is normally denoted by $\mathbb{N}$. Thus the set $\mathbb{N}$ of natural numbers consists of the numbers $0, 1, 2, 3, 4, \ldots$.

Over the centuries various ways have been devised for recording these numbers. The actual symbols we now use for them—the so-called arabic numerals—originated in India many centuries ago. In time, they were incorporated into Arab culture and were then finally brought to Europe by the Arabs. Ancient civilisations, however, commonly used marks of various shapes to denote the number of objects, with each mark corresponding to one object in much the same way as primitive tribesmen sometimes make notches on the handles of their weapons to show how many of the enemy they have slain.

It is, of course, hardly surprising that it soon became recognised that counting alone was not sufficient and the laws for the operations of addition and multiplication were developed together with rules governing inequalities. Thus two numbers $m, n$ could be compared and statements of the form $m < n$ or $m \leqslant n$ have sensible meanings. It is the order properties of the natural numbers which gives rise to the powerful method of proof called the principle of mathematical induction. Proofs using this principle permeate most branches of mathematics and an appendix at the end of the chapter discusses it.

Many devices like the abacus were developed to facilitate addition and multiplication and their users became amazingly skilled and displayed great ingenuity in handling them—though they were naturally somewhat slower than our modern computers! But at least users of the abacus obtained the answer 4 for $2 \times 2$ and not 3.999. On the whole, the abacus was a commercial tool. It was used widely by merchants and traders. The highly educated, however, tended to prefer written records and some of the mathematics of the Babylonians, Egyptians, Greeks, etc. can be seen in museums today.

If $m$ and $n$ are two natural numbers, then their sum $m+n$ and their product $mn$ are also natural numbers. However, given two natural numbers $m, n$ it is not always possible to find a natural number $x$ such that

$$m + x = n. \tag{1}$$

For instance, there is no natural number $x$ such that $3 + x = 2$. In order to be able to solve equation (1) to find $x$ in every case, the number system must be extended to the set $\mathbb{Z}$ of all integers.

## 1.2 INTEGERS

The set $\mathbb{Z}$ of integers consists of the numbers $0, \pm 1, \pm 2, \pm 3, \pm 4, \ldots$. The set of positive integers which forms part of this set is denoted by $\mathbb{Z}^+$. Moreover, given any two integers $m, n$ it is always possible to find an integer $x$ such that

$$m + x = n$$

and such equations no longer cause any difficulty. As in the case of the natural numbers an order relation is defined on $\mathbb{Z}$, so that statements of the form $n \leqslant p$ or $p < q$ have a sensible meaning.

Suppose we have a non-empty set $S$ of integers with the property that there is some integer $p$ such that

$$p \leqslant m \tag{1}$$

for all $m \in S$. Then we can easily show that the set $S$ possesses a smallest element. This result is known as **the well-ordering principle** for the integers.

Since $S$ is not empty, it must contain some element $q$. Let the set $T$ consist of all the integers $l$ such that $p \leqslant l \leqslant q$,

i.e. $$T = \{p, p+1, p+2, p+3, \ldots, q-1, q\}.$$

Then $T$ has only a finite number of elements, and therefore $T \cap S$ has only a finite number of elements. Moreover, $T \cap S$ is not empty, because $q \in T$ and $q \in S$; and $T \cap S$ contains all the elements of $S$ which are either less than $q$ or equal to $q$. Since $T \cap S$ has only a finite number of elements it must have a smallest element $k$. This integer $k$ belongs to $S$ and there are no elements of $S$ which are less than $k$, i.e. $k$ is the smallest element of $S$.

Since every positive integer satisfies the inequality $m \geqslant 1$ (which is a special case of (1) with $p = 1$), we see that every non-empty set of positive integers has a least element. This is precisely the property we need to prove the principle of mathematical induction.

We record these results for future reference in the form of a theorem and its corollary.

THEOREM 1.2.1 Let $S$ be a non-empty set of integers with the property that there is some integer $p$ such that $p \leqslant m$ for all $m \in S$. Then $S$ has a least member.

COROLLARY Every non-empty set of positive integers has a least member.

We now recall that the sum and product of two integers is always an integer. However, given two integers $p, q$ with $p \neq 0$, it is not always possible to find an integer $x$ such that

$$px = q.$$

For example, there is no integer $x$ such that $3x = 2$. This hurdle can be surmounted only by again extending the number system. In this case we obtain the set $\mathbb{Q}$ of rational numbers.

## 1.3 RATIONAL NUMBERS

The set of rational numbers consists of all numbers of the form $p/q$, where $p, q$ are integers and $q \neq 0$. For example $\frac{1}{2}$, $-\frac{2}{3}$, $\frac{27}{11}$, etc. are all rational numbers. The use of the rational numbers permits the solution of the equations

$$r + x = s,$$

$$tx = u$$

for $x$, where $r, s, t, u$ are all rational numbers and $t \neq 0$. Thus operations of subtraction and division (with non-zero divisor) are always possible with rational numbers.

Any rational number can be represented as the quotient of two integers or as a decimal in the familiar way. But, does this decimal have any special features? Yes, it does. The decimal representation of a rational number is either terminating, i.e. it has only a finite number of digits after the decimal point, or it is a repeating decimal, i.e. after a certain time the digits keep on repeating a definite pattern. For example $\frac{1}{8} = 0.125$, $\frac{1}{7} = 0.142\,857\,142\,857\,1\ldots$ with the pattern 142 857 continually repeating. Normally we put dots over the repeating pattern of digits and we would write $\frac{1}{7} = 0.\dot{1}\dot{4}\dot{2}\,\dot{8}\dot{5}\dot{7}$, $\frac{39}{11} = 3.\dot{5}\dot{4}$ (which means that $\frac{39}{11} = 3.545\,454\,545\,4\ldots$) and $\frac{5}{6} = 0.8\dot{3}$ (which means that $\frac{5}{6} = 0.833\,333\,333\ldots$). In fact we can also prove the converse, viz. that terminating decimals and repeating decimals always represent rational numbers. The reader should try calculating the decimal representation of some rational numbers; this will provide insight into the reason why these patterns occur.

Nowadays, pocket calculators are frequently used to find the quotient of two numbers. If, however, we recall the old method of long division and use it to divide $p$ by $q$, where $p$, $q$ are integers and $q > 0$, then the reason for the repeating pattern becomes immediately obvious. First let us suppose that $0 < p < q$ and let us use the method of long division to express $p/q$ as a decimal. At each stage we divide by $q$ to give a remainder $r$ which is one of the numbers $0, 1, 2, 3, \ldots, (q-1)$. Then we bring down the digit 0 and divide $r \times 10$ by $q$, to give a remainder and keep on repeating this process. Since there are only $q$ different possible values for the remainder, we either obtain zero remainder at some stage and the decimal terminates, or, after at most $q$ steps of long division, we obtain a remainder which is the same as one of the previous remainders and the pattern of digits starts to repeat. The reader is encouraged to try some examples to check what happens.

If $0 < q < p$ then $\frac{p}{q}$ can be expressed as $\frac{p}{q} = m + \frac{n}{q}$, where $m, n$ are integers and $0 \leqslant n < q$. Thus $n/q$ is represented by a decimal which either terminates or has a repeating pattern, hence it follows that the decimal representation of $p/q$ either terminates or has a repeating pattern. The extension of the argument to cover the case in which the rational number $p/q$ is negative is now obvious.

Conversely, it can be proved that a decimal with a repeating pattern represents a rational number, and a terminating decimal obviously represents a rational number. The proof is left as an exercise for the reader.

## *EXERCISE 1.3.1*

First consider a decimal of the form $0.\dot{a}_1\dot{a}_2\ldots\dot{a}_n$ with a repeating pattern of $n$ digits. Write $x = 0.\dot{a}_1\dot{a}_2\ldots\dot{a}_n$. Express $10^n x$ as a decimal. Then subtract $x$ and check that $10^n x - x$ is an integer. Deduce that $x$ is rational. Now extend the method to prove that all decimals with a repeating pattern represent rational numbers.

It is interesting to consider how many rational numbers there are. Obviously there are infinitely many, but we can get some idea of the size of $\mathbb{Q}$ by noticing that its elements can be listed. To prepare this list we first notice that the positive rationals can be displayed in a doubly infinite array as follows:

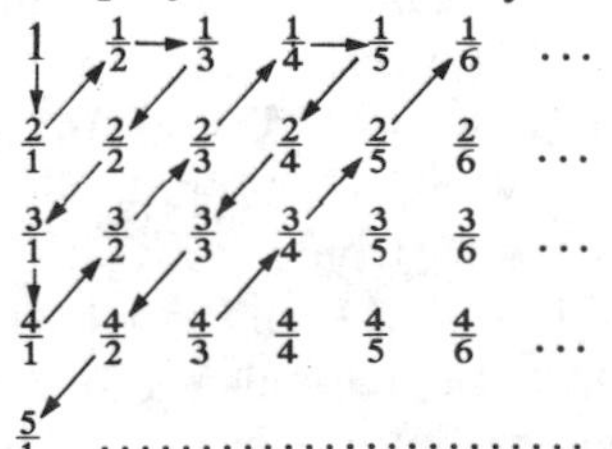

This can now be reduced to a single list by starting at the top left-hand corner and proceeding along the diagonals as shown by the arrows. Each time we come to a number which has already been included, we pass over it and continue on our way following the arrows. This gives a list beginning $1, 2, \frac{1}{2}, \frac{1}{3}, 3, 4, \frac{3}{2}, \frac{2}{3}, \frac{1}{4}, \frac{1}{5}, 5, \ldots$. Now that the positive rationals are arranged in a certain order, we include all the remaining rationals by inserting 0 at the beginning and then inserting the negative element $-p/q$ directly after $p/q$ in the list. We have now included all the rationals. Clearly there are infinitely many of them, but not so many that they cannot be incorporated in a single list. We will see later that the real numbers are too numerous to be put in order in a list. We say that the set $\mathbb{Q}$ of rationals is countably infinite. But are there enough of them for the purpose of analysis?

At the beginning of the chapter we started with the familar set of natural numbers and it rapidly became apparent that this number system had to be extended with the result that we now have the rational number system $\mathbb{Q}$. Is this adequate or do we need further extension? Certainly, many numerical calculations use only rational numbers. A modern computer can use numbers with a very large number of digits but its capacity is finite and so any decimal it uses must have only a finite number of digits and it is, therefore, a rational number. If a calculation requires a number whose square is 2, a computer or pocket calculator will use a rational number for it. Is this justified? Is there a rational number whose square is 2?

Centuries before Christ, the Babylonians had tables of roots. A surviving tablet shows that they gave

$$1 + \frac{24}{60} + \frac{51}{(60)^2} + \frac{10}{(60)^3}$$

as the value of the square root of 2 (in our decimal system this is 1.414 212 96 which is in remarkable agreement with the value given in our four-figure tables). How they calculated their roots is not known, but their results are surprisingly accurate. Obviously, they expressed their values in the form

$$p_0 + \frac{p_1}{60} + \frac{p_2}{(60)^2} + \frac{p_3}{(60)^3} + \frac{p_4}{(60)^4} + \cdots,$$

where $p_0, p_1, p_2, \ldots$ are integers and they used more and more terms to obtain closer approximations. We imagine they believed that the process would terminate eventually giving a finite number of terms for the square root of 2 and so assumed that there is a rational number whose square is 2. Surviving records show that the Greeks and Egyptians also knew how to calculate good approximations to roots. The Greeks, of course, concentrated on the geometric aspects of mathematics. Using a ruler and compass they could construct a unit square and they knew that the area of the square on its diagonal was 2. So what is the length of this diagonal?

A leading Greek mathematician Pythagoras (who lived in the sixth century BC) gathered together a group of disciples and founded the Pythagorean Brotherhood which lasted probably two hundred years. They had strict rules for the conduct of their life and made known most of their mathematical discoveries as a corporate endeavour. The Pythagoreans were able to prove that there is no unit of length which can be used for measuring for which the sides of the square and the diagonal are all whole numbers of units, i.e. they recognised that the ratio of the length of the diagonal to the length of the side of the square could not be expressed as the ratio of two natural numbers. Thus there are no natural numbers $p, q$ such that $(p/q)^2 = 2$. In Pythagorean terminology this meant that these two lengths were incommensurable. Thus the square root of 2 exists as a length in geometry but it is not a rational number. A quantity like this which cannot be expressed as the ratio of two natural numbers was not acceptable to the Pythagoreans as a number at all; it was called a magnitude and a distinction was drawn between numbers and magnitudes. Today of course we would refer to Pythagorean magnitudes as irrational numbers.

The strict adherence to the principle that the only numbers are those which can be expressed as the ratio of two natural numbers gave rise in time to many paradoxes. Perhaps the most famous were Zeno's paradoxes and the well-known paradox of Achilles and the tortoise. Two centuries later the situation was eased by the introduction of definitions and theorems about ratios of magnitudes.

For us today, perhaps the simplest way of proving that there is no rational number whose square is 2 is to use a contradiction argument. This method of proof was invented by the Greeks and has become accepted as a traditional method of proof. It begins by assuming that the result is false and shows that this leads logically to a contradiction. Let us suppose that there is a rational number $p/q$ whose square is 2, where $p, q$ are positive integers with no common factor. Then

$$\left(\frac{p}{q}\right)^2 = 2,$$

and therefore

$$p^2 = 2q^2. \tag{1}$$

Hence $p^2$ has a factor 2 and so $p$ is even. Put $p = 2r$, where $r$ is an integer in (1), to obtain

$$4r^2 = 2q^2, \qquad \text{i.e.} \qquad q^2 = 2r^2.$$

Thus $q^2$ has a factor 2 and so $q$ is also even, which means that $p, q$ both have a factor 2. This contradicts the assumption that $p, q$ have no common factors. Thus the initial assumption that there is a rational number $p/q$ whose square is 2 leads to a contradiction. This assumption is therefore false, i.e. there is no rational number whose square is 2 and there is a glaring gap in the rational numbers. Whether the Greeks used this method to prove their conjecture is not known. It is possible they used an argument along the same lines, as they stated that they used a method depending on even and odd natural numbers. It is easy to show that there are many other such gaps in the rationals. This, certainly, has rather far-reaching consequences. Perhaps one simple example gives some idea of the problems encountered because of the gaps in the rationals.

Consider a very simple function given by $f(x)=x^2$ ($x \in \mathbb{Q}$). Then $f(1)=1$ and $f(2)=4$. If we draw the graph of this function then we normally take it for granted that $f(x)$ assumes every value between 1 and 4 as $x$ increases from 1 to 2. However, if we only use rational values of $x$ between 1 and 2, then $f(x)$ cannot take the value 2, because there is no rational whose square is 2. Thus $f(x)$ does not take all rational values between 1 and 4 as $x$ takes all rational values between 1 and 2. This illustrates how our intuitive ideas about continuous functions fall apart when only rational numbers are used. We must therefore extend our number system. What we need for the purposes of analysis is the real number system.

## 1.4 THE REAL NUMBER SYSTEM

By the fourth century BC the Greeks had seen the need for definitions and theorems for ratios of magnitudes. Because of their preoccupation with geometry and in particular their interest in right-angled triangles, the magnitudes which probably interested them most were square roots of natural numbers like 2, 5, etc. These square roots are, of course, only very special examples of the numbers we call irrational numbers today. Admittedly they used the letter $\pi$ for the ratio of the circumference of a circle to its diameter. But we are not certain whether they knew that it could not be expressed as the ratio of two natural numbers.

The Greeks had a preoccupation with whole numbers (i.e. natural numbers). If they wished to consider the ratio of two lengths, they tried to find some sufficiently small unit of length so that the two given lengths could both be expressed as a whole number of units. When they compared the diagonal of a square with the sides, their search for a unit of measurement led them to keep on continually dividing to find smaller and smaller units in the vain hope of expressing each length as a whole number of units. Thus, when theorems were devised for ratios of magnitudes, the Greeks had to introduce the idea of infinite divisibility. Ideas of infinity were foreign to Greek mathematics and they preferred to avoid them, and to use rational numbers. Many centuries elapsed before a satisfactory theory of real numbers and of infinite sets began to emerge.

Probably the most significant advances in these areas began in the second half of the nineteenth century with the work of five mathematicians—four of them Germans and one a Frenchman. Their work provided the foundation for much of the analysis we know today and their names have lived on in many theorems. They were Karl Weierstrass (1815–97), Eduard Heine (1821–81), Georg Cantor (1845–1918), Richard Dedekind (1831–1916) and Charles Méray (1835–97), and they set about the task of defining the phrase 'real number'. The algebraic and order properties of real numbers had long been known and raised no great problems. For the nineteenth-century mathematician the real task was how to express in mathematical symbols the idea that the real numbers form a continuum—a property which immediately distinguishes them from the rational numbers. This property is embodied in an axiom which we call the completeness axiom for the real number system. It ensures that there are no gaps in the real number system. For example, we can use it to show that there is a positive real number whose square is 2.

The completeness axiom can be stated in many equivalent forms. We choose the formulation in terms of the 'supremum' (i.e. the least upper bound) because it seems the most convenient form to handle in the present context.

We do not set about constructing the real numbers from the rationals. Instead, we take the viewpoint that the real numbers are the familiar objects which we have used throughout our time at secondary school and we list a set of axioms which they must satisfy. These axioms distinguish the real number system from other number systems (i.e. any number system which satisfies them is mathematically indistinguishable from the real number system). We use the letter $\mathbb{R}$ for the set of all real numbers. The axioms (or basic rules) which must be satisfied fall broadly into three groups. The first group governs the way in which the familiar operations of addition and multiplication are carried out.

## Algebraic axioms

**A1** For all $x, y \in \mathbb{R}$, $x + y \in \mathbb{R}$ and $xy \in \mathbb{R}$.

**A2** For all $x, y \in \mathbb{R}$,

$$(x+y)+z = x+(y+z). \qquad \text{(associative law)}$$

**A3** For all $x, y \in \mathbb{R}$,

$$x+y = y+x. \qquad \text{(commutative law)}$$

**A4** There is a number $0 \in \mathbb{R}$ such that

$$x+0 = x = 0+x \qquad \text{for all } x \in \mathbb{R}.$$

**A5** For each $x \in \mathbb{R}$, there exists a corresponding number $(-x) \in \mathbb{R}$ such that

$$x+(-x) = 0 = (-x)+x.$$

**A6** For all $x, y, z \in \mathbb{R}$,

$$(xy)z = x(yz). \qquad \text{(associative law)}$$

**A7** For all $x, y \in \mathbb{R}$,

$$xy = yx. \qquad \text{(commutative law)}$$

**A8** There is a number $1 \in \mathbb{R}$ such that

$$x \cdot 1 = x = 1 \cdot x \qquad \text{for all } x \in \mathbb{R}.$$

**A9** For each $x \in \mathbb{R}$ such that $x \neq 0$, there is a corresponding number $(x^{-1}) \in \mathbb{R}$ such that

$$x(x^{-1}) = 1 = (x^{-1})x.$$

**A10** For all $x, y, z \in \mathbb{R}$

$$x(y+z) = xy + xz. \qquad \text{(distributive law)}$$

The second group of axioms governs the way we use inequalities.

### Axioms for order relations

**O1** Any pair $x, y$ of real numbers satisfies precisely one of the following relations:

(a) $x < y$; (b) $x = y$; (c) $y < x$.

**O2** If $x < y$ and $y < z$ then $x < z$.
**O3** If $x < y$ then $x + z < y + z$.
**O4** If $x < y$ and $z > 0$ then $xz < yz$.

These two groups of axioms are given for reference. Their relevance is not, of course, restricted to the real number system as the rational numbers also satisfy them. It is the next axiom (which embodies the idea that the real numbers form a continuum) which underpins much of our analysis. Since the completeness axiom is stated in terms of upper bounds, some preliminary definitions are needed before the axiom can be formulated at all.

DEFINITION 1.4.1 A set $S$ of real numbers is said to be **bounded above** if there is some number $M$ such that $x \leqslant M$ for **all** $x \in S$. The number $M$ is called **an upper bound** for $S$.

DEFINITION 1.4.2 A set $S$ of real numbers is said to be **bounded below** if there is some real number $m$ such that $x \geqslant m$ for **all** $x \in S$. The number $m$ is called **a lower bound** for $S$.

DEFINITION 1.4.3 A set $S$ of real numbers is said to be **bounded** if there is some real number $k$ such that $|x| \leqslant k$ for **all** $x \in S$.

If the set $S$ is bounded then $|x| \leqslant k$ for all $x \in S$ and some real number $k$. It follows that $-k \leqslant x \leqslant k$ for all $x \in S$ and so $S$ is bounded above and also bounded below, i.e. a bounded set is bounded above and bounded below.

---

*Examples 1.4.1*

1. Let $S$ consist of all the real numbers $x$ such that $0 \leqslant x \leqslant 1$. We shall write this in the standard way as

$$S = \{x \in \mathbb{R} : 0 \leqslant x \leqslant 1\},$$

i.e. $S$ consists of all the $x \in \mathbb{R}$ such that $0 \leqslant x \leqslant 1$. Then $S$ is bounded above and 1 is an upper bound. Any number $M \geqslant 1$ is also an upper bound for $x \leqslant 1 \leqslant M$ for all $x \in S$. Thus $S$ has an infinite number of upper bounds. Clearly $S$ is also bounded below and 0 is a lower bound. Any number $m \leqslant 0$ is also a lower bound as $m \leqslant 0 \leqslant x$ for all $x \in S$. Moreover, $|x| \leqslant 1$ for all $x \in S$ and $S$ is bounded.

2. Let $S = \{\frac{1}{n} : n \in \mathbb{Z}^+\}$, i.e. $S$ consists of the numbers $1, \frac{1}{2}, \frac{1}{3}, \frac{1}{4}, \ldots$. Then $S$ is bounded above and 1 is an upper bound (in fact any number $M \geqslant 1$ is also an upper bound), $S$ is bounded below and 0 is a lower bound (any number $m \leqslant 0$ is also a lower bound). Moreover, since $|x| \leqslant 1$ for all $x \in S$, $S$ is also bounded.
3. Let $S = \{x \in \mathbb{R} : x > 1\}$. Then $S$ is not bounded above but it is bounded below and any number $m \leqslant 1$ is a lower bound.
4. Let $S = \{x \in \mathbb{R} : 0 < x < 1\}$, then $S$ is bounded above and any number $M \geqslant 1$ is an upper bound. It is also bounded below and any number $m \leqslant 0$ is a lower bound.

---

The set $S$ in Example 4 has no least member, despite the fact that it is bounded below. However its set of upper bounds has a least member, and, furthermore, the sets in Examples 1 and 2 have the same property. This least element of the set of upper bounds is called the **supremum**, i.e. **the supremum is the least upper bound**. We also see that the sets in Examples 1 to 4 are bounded below and the set of lower bounds has a greatest member. This greatest member of the set of lower bounds is called the **infimum**, i.e. **the infimum is the greatest lower bound**.

---

*Examples 1.4.2*

1. The set $S = \{x \in \mathbb{R} : 0 \leqslant x \leqslant 1\}$ has supremum 1 and infimum 0 and both the supremum and infimum are elements of the set.
2. The set $S = \{\frac{1}{n} : n \in \mathbb{Z}^+\}$ has supremum 1 and infimum 0 and the supremum is an element of $S$. The infimum, however, does not belong to $S$.
3. The set $S = \{x \in \mathbb{R} : x > 1\}$ is not bounded above and does not have a supremum. It is bounded below and its infimum is 1, which does not belong to the set.
4. The set $S = \{x \in \mathbb{R} : 0 < x < 1\}$ has supremum 1 and infimum 0 and neither the supremum nor the infimum belong to $S$.

---

Perhaps the reader will find it instructive to spend time having a closer look at some or all of the previous examples. For instance, in Example 4, the reader will notice that 1 is certainly an upper bound for $S$, since $x < 1$ whenever $x \in S$. Moreover, any number less than 1 is not an upper bound. In fact, if $\varepsilon > 0$ is any positive real number (however small), then $1 - \varepsilon$ is not an upper bound, since there are elements of $S$ between $1 - \varepsilon$ and 1. Hence 1 is the least member of the set of upper bounds and so it is the supremum of $S$. These ideas lead to a very useful characterisation of the supremum. Suppose the set $S$ has supremum $\alpha$; then (a) $\alpha$ is an upper bound of $S$, i.e. $x \leqslant \alpha$ for all $x \in S$ and (b) any number less than $\alpha$ is not an upper bound; i.e., given any $\varepsilon > 0$ (however small), $\alpha - \varepsilon$ is not an upper bound of the set. There is, therefore, some element $x' \in S$ such that $\alpha - \varepsilon < x' \leqslant \alpha$. These ideas prove so valuable in practice that they are worth incorporating in a theorem.

THEOREM 1.4.1 Let $S$ be a non-empty set of real numbers. Then the real number $\alpha$ is the supremum of $S$ if and only if both the following conditions are satisfied:

(a) $x \leqslant \alpha$ for **all** $x \in S$;
(b) for every $\varepsilon > 0$, there is some $x' \in S$ such that $\alpha - \varepsilon < x' \leqslant \alpha$.

*Proof* We notice that condition (a) ensures that $\alpha$ is an upper bound of $S$ and condition (b) guarantees that any number $\alpha - \varepsilon$ (which is less than $\alpha$) is not an upper bound. Hence $\alpha$ is the least of the upper bounds. It follows that if $\alpha$ satisfies (a) and (b) then $\alpha$ is the supremum of $S$. Conversely, if $\alpha$ is the supremum of $S$, then $\alpha$ must satisfy (a) and (b).

A similar result holds for the infimum.

THEOREM 1.4.2 Let $S$ be a non-empty set of real numbers. Then the real number $\beta$ is the infimum of $S$ if and only if both the following conditions are satisfied:

(a) $\beta \leqslant x$ for **all** $x \in S$,
(b) for every $\varepsilon > 0$, there is some $x' \in S$ such that $\beta \leqslant x' < \beta + \varepsilon$.

The first condition, $\beta \leqslant x$ for all $x \in S$, ensures that $\beta$ is a lower bound of $S$ and the second condition guarantees that any number $\beta + \varepsilon$ (which is greater than $\beta$) is not a lower bound. Thus $\beta$ is the greatest of the lower bounds.

These results for the supremum and infimum may at first sight seem trivially obvious and not particularly worthy of such emphasis. They will, however, prove to be of great value later.

The supremum (and to a lesser extent the infimum) plays a central role in the development of the real number system. For this reason it is convenient to have some abbreviation of the notation. Normally the supremum of a set $S$ is denoted by either $\sup S$ or $\sup_{x \in S} x$. The infimum of $S$ is denoted by either $\inf S$

or $\inf_{x\in S} x$. In the cases in which either $\sup S$ or $\inf S$ exist they may be elements of $S$, but they do not necessarily belong to $S$.

## *EXERCISES 1.4.1*

**1** In each of the following cases, decide whether the set $S$ has (i) a supremum or (ii) an infimum:

(a) $S=\{x\in\mathbb{R}: a\leqslant x\leqslant b\}$;
(b) $S=\{x\in\mathbb{R}: a<x<b\}$;
(c) $S=\left\{1+(-1)^n\frac{1}{n}: n\in\mathbb{Z}^+\right\}$;
(d) $S=\left\{n+(-1)^n\frac{1}{n}: n\in\mathbb{Z}^+\right\}$;
(e) $S=\{1+(-1)^n n: n\in\mathbb{Z}^+\}$;
(f) $S=\{x\in\mathbb{R}: 3x^2-10x+3<0\}$.

(Remember, we have used $\mathbb{Z}^+$ for the set of positive integers.)

In each case in which either the supremum or infimum exists, decide whether they are elements of $S$.

**2** Let $A, B$ be two non-empty sets of real numbers with supremums $\alpha$ and $\beta$ respectively, and let the sets $A+B$ and $AB$ be defined by

$$A+B=\{a+b: a\in A, b\in B\},$$

$$AB=\{ab: a\in A, b\in B\}.$$

Show that $\alpha+\beta$ is the supremum of $A+B$. Give an example to show that $AB$ need not have a supremum. Prove also that even if $AB$ has a supremum, this supremum need not be equal to $\alpha\beta$.

Now that the reader has acquired a certain familiarity with the idea of the supremum, we use it to formulate the completeness axiom for the real numbers.

COMPLETENESS AXIOM **Every non-empty set of real numbers which is bounded above has a supremum.**

Later, we see that the requirement that the real number system satisfies the completeness axiom does indeed guarantee that the real numbers form a continuum and ensures that we have sufficient numbers for the purposes of analysis.

In a way, the completeness axiom may at first appear a little lopsided. It is formulated in terms of upper bounds and the supremum. Lower bounds do not seem to even merit a mention. However, appearances can at times be deceptive! The completeness axiom ensures that non-empty sets which are bounded below have an infimum as the next theorem shows. The completeness axiom itself is not a theorem which has to be proved. It is a basic property which a number system is required to possess if it is to be called the real number system. Our next theorem has a different standing. It is a deduction made on the understanding that real numbers, by definition, satisfy the completeness axiom.

THEOREM 1.4.3 A non-empty set of real numbers which is bounded below has an infimum.

*Proof* Let $S$ be a non-empty set which is bounded below and let $S^*$ consist of all the numbers of the form $-x$, where $x \in S$.

i.e. $$S^* = \{-x : x \in S\}.$$

Since $S$ is bounded below, it has a lower bound $m$ and

$$m \leqslant x \qquad \text{for all } x \in S.$$

Hence $$-m \geqslant -x \qquad \text{for all } x \in S.$$

Thus $-m$ is an upper bound for $S^*$. Since $S^*$ is non-empty and bounded above it has a supremum $\alpha$ by the completeness axiom. Hence, using Theorem 1.4.1, we see that

(a) every element of $S^*$ is less than or equal to $\alpha$,

i.e. $$-x \leqslant \alpha \qquad \text{for all } x \in S,$$

and

(b) for any $\varepsilon > 0$, there is an element $-x'$ of $S^*$ with $\alpha - \varepsilon < -x' \leqslant \alpha$

i.e. $$-\alpha + \varepsilon > x' \geqslant -\alpha \qquad \text{for some } x' \in S.$$

The above statements show that $-\alpha$ is the infimum of $S$.

The remainder of the chapter will be devoted to consequences of the completeness axiom. There are many properties of real numbers which the reader may regard as obvious and never really question which, surprisingly, require the completeness axiom for their proof. As the first illustration of this we consider the Archimedean property which states that if $x$ is any real number then there is an integer greater than $x$. Of course, the result is trivially true for $x$ negative. For if $x < 0$ then zero is an integer exceeding $x$. It is the proof for $x > 0$ which requires the completeness axiom.

THEOREM 1.4.4 (Archimedean Property) Given any real number $x$, there is an integer $p$ such that $p > x$.

*Proof* We employ the traditional contradiction argument which we met earlier in the chapter. If the result is not true, then there is some real number $x^*$ such that no integer exists greater than $x^*$, i.e. $p \leqslant x^*$ for all integers $p$. Hence the set $\mathbb{Z}$ of all integers is non-empty and bounded above by $x^*$. By the completeness axiom, $\mathbb{Z}$ has a supremum $\alpha$, and therefore $p \leqslant \alpha$ for all integers $p$. Now, if $p$ is an integer then $p+1$ is also an integer and so

$$p+1 \leqslant \alpha \tag{1}$$

(since all integers are less than or equal to $\alpha$). Hence, from (1), it follows that

$$p \leqslant \alpha - 1.$$

Since this is true for all integers $p$, we see that $\alpha - 1$ is an upper bound for $\mathbb{Z}$, which contradicts the fact that $\alpha$ is the supremum of $\mathbb{Z}$. We therefore have the necessary contradiction. This contradiction follows as a logical consequence of the assumption that the result is false. Thus the proof is complete.

COROLLARY Given any real number $x$, there is an integer $n$ such that $n < x$.

*Proof* We first note that $-x$ is a real number whenever $x$ is real. By the Archimedean property, there is an integer $p > -x$ and therefore $-p < x$. Now $-p$ is an integer whenever $p$ is an integer and the proof is complete.

These properties are certainly of interest to the analyst, but what about our previous claim that the completeness axiom ensures that the real numbers form a continuum and there are no gaps? We made a promise that we could show, for example, that there is a real number whose square is 2. Now is the time to carry it out. The details of the proof are rather tedious, but the result itself is exciting.

THEOREM 1.4.5 There is a positive real number $c$ such that $c^2 = 2$.

*Proof* Let the set $S$ consist of all the positive real numbers $x$ such that $x^2 < 2$.

i.e. $$S = \{x \in \mathbb{R} : x > 0 \text{ and } x^2 < 2\}.$$

Then $1 \in S$ and $S$ is not empty. Moreover, if $x > 2$, then $x^2 > 4$ and $x \notin S$. Hence $x \leqslant 2$ for all $x \in S$ and $S$ is bounded above. By the completeness axiom $S$ has a supremum $c$. We will now show that $c^2 = 2$, by proving (a) $c^2 \nless 2$ and (b) $c^2 \ngtr 2$.

(a) *We prove* $c^2 \nless 2$ This is proved using a contradiction argument. For suppose the result is false, i.e. $c^2 < 2$, and write

$$h = \frac{2 - c^2}{4} = \frac{1}{2} - \frac{c^2}{4}. \tag{1}$$

Then, clearly, $$0 < h < \tfrac{1}{2},$$

and therefore $$c < \frac{c}{1-h}. \tag{2}$$

But from (1),

$$\left(\frac{c}{1-h}\right)^2 = \frac{c^2}{1 - 2h + h^2} < \frac{c^2}{1 - 2h} = \frac{c^2}{1 - \left(1 - \dfrac{c^2}{2}\right)} = 2.$$

i.e. $$\left(\frac{c}{1-h}\right)^2 < 2$$

and $[c/(1-h)] \in S$. Since, from (2), $[c/(1-h)] > c =$ supremum of $S$, this gives the required contradiction and we cannot possibly have $c^2 < 2$.

(b) *We prove* $c^2 \ngtr 2$ Again a traditional contradiction argument is used. For suppose the result is false, i.e. suppose $c^2 > 2$ and write

$$k = \frac{c^2 - 2}{2c^2} = \frac{1}{2} - \frac{1}{c^2}. \tag{3}$$

Then, clearly, $$0 < k < \tfrac{1}{2},$$

and therefore $$0 < c(1-k) < c. \tag{4}$$

Moreover,

$$[c(1-k)]^2 = c^2(1 - 2k + k^2) > c^2(1 - 2k) = c^2\left(1 - \left\{1 - \frac{2}{c^2}\right\}\right) = 2.$$

Now if $x \in S$, then $x^2 < 2 < [c(1-k)]^2$, and therefore $0 < x < c(1-k)$ whenever $x \in S$. Thus $c(1-k)$ is an upper bound for $S$, and, from (4), $c(1-k) < c = \sup S$, which gives the needed contradiction. Thus the assumption $c^2 > 2$ cannot hold.

Since $c^2 \nless 2$ and $c^2 \ngtr 2$ and $c^2$ exists we must have $c^2 = 2$. This proves that there is a positive real number whose square is 2. Normally we use $\sqrt{2}$ for this number. We have already proved that it is not a rational number and so $\sqrt{2}$ is a real number which is not rational. Such a number is called an irrational number.

DEFINITION 1.4.4 A real number which is not a rational number is called an **irrational** number.

The irrational numbers are the numbers belonging to $\mathbb{R}\backslash\mathbb{Q}$, i.e. the numbers which belong to $\mathbb{R}$ but not to $\mathbb{Q}$. The number $\sqrt{2}$ is an example of an irrational number. Other well-known examples of irrational numbers are $\pi$ and e. Lambert gave the first proof of the irrationality of $\pi$ in a paper to the Berlin Academy in 1761. Essentially his argument was quite simple. He showed that if $x$ is a non-zero rational number, then $\tan x$ is irrational. Since $\tan(\pi/4) = 1$ which is rational, the number $\pi/4$ cannot be rational, i.e. $\pi$ is not rational. Subsequently Legendre proved the irrationality of $\pi$ independently in 1794. The irrationality of e had already been established by Euler in 1737.

Suppose we now take $S^*$ to be the set of all positive rational numbers $x$ such that $x^2 < 2$, then $S^*$ is not empty and it is bounded above. If we take $A$ to be the set of all rational numbers $a$ which are upper bounds of $S^*$, then we will show later that $A$ does not have a least member, which demonstrates effectively that the set $\mathbb{Q}$ of rational numbers does not satisfy the completeness axiom. The completeness axiom, therefore, plainly distinguishes between the real number system and the rational number system.

## *EXERCISES 1.4.2*

**1** Prove that if $x^2$ is irrational then $x$ must be irrational. Is the converse true?

**2** Show that if $x$ is irrational and $y$ is rational then $x + y$ is irrational.

**3** If $x$ is irrational and $y$ is irrational, is it true that $x + y$ is irrational? Give reasons for your answer. Is $xy$ irrational?

We have already seen that early Greek mathematicians, living centuries before the birth of Christ, knew of the existence of the irrational number $\sqrt{2}$ as a length in geometry. Their ideas suggest an obvious geometric representation for real numbers.

Draw a straight line from left to right across the page. Now select a point on it (called the origin) and label it with the number 0. Then choose a unit of length and label points to the right of 0 which are $1, 2, 3, \ldots$ units of length from 0 with the numbers $1, 2, 3, \ldots$. Mark points to the left of 0, which are distances $1, 2, 3, \ldots$ from 0 and label them $-1, -2, -3, \ldots$ respectively (see Fig. 1.1). Any point $P$ to the right of 0 on the line represents the real number $x$, where $x$ is the length of $0P$. Any point $P'$ to the left of 0 on the line represents the real number $-x'$, where $x'$ is the length of $0P'$. Thus every point on the line represents a unique real number. Conversely, given any real

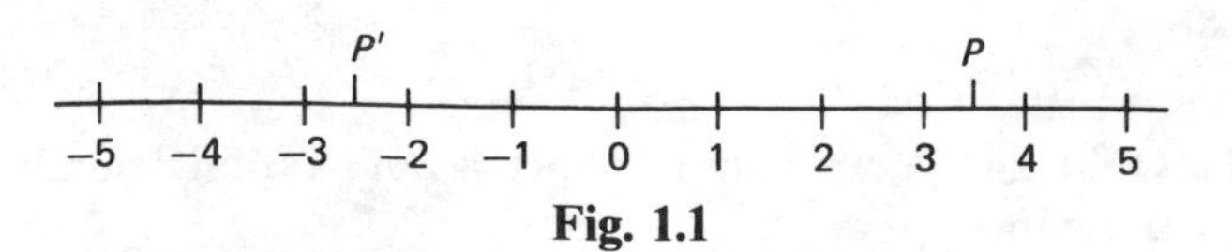


**Fig. 1.1**

number there is one and only one corresponding point on the line. This, of course, represents pictorially the idea that the real numbers form a continuum.

Now suppose we take our picture and on the line we initially mark only the points corresponding to rational numbers. Then, certainly, there will be some holes. For example, there will be holes at $\sqrt{2}, \sqrt{3}, \sqrt{5}, \ldots, \mathrm{e}, \pi, \mathrm{e}^2, \pi^2$, etc.

But how much of the line will be missing and how many holes will there be?

If we mark only the points representing rational numbers, then there will be 'more holes' than points marked. Without going into detail as to what precisely is meant by 'more holes', we can justify this statement by proving that there are so many real numbers that they cannot be arranged in a list. In fact, there are so many real numbers between 0 and 1 that they cannot be so arranged. However hard we try and however careful we are in compiling the list, we will always find that there are so many numbers that some have been omitted.

Again we use a contradiction argument. Just suppose we can list the real numbers between 0 and 1 in their decimal representation as

$$
\begin{gathered}
0.a_{11}a_{12}a_{13}a_{14}\ldots \\
0.a_{21}a_{22}a_{23}a_{24}\ldots \\
0.a_{31}a_{32}a_{33}a_{34}\ldots \\
0.a_{41}a_{42}a_{43}a_{44}\ldots \\
\ldots\ldots\ldots\ldots\ldots\ldots\ldots,
\end{gathered}
$$

and recall that a real number has a unique decimal expansion if we exclude decimals with recurring nines. Consider the real number $x = 0.b_1b_2b_3b_4\ldots$, where $b_k = 4$ if $0 \leqslant a_{kk} \leqslant 2$ and $b_k = 1$ if $a_{kk} \geqslant 3$. Then $x$ cannot be equal to the first number in the list because of the way in which $b_1$ and $a_{11}$ differ. Similarly, $x$ differs from the second number in the list and so on. Hence $x$ is not in the list, despite the fact that $0 < x < 1$. This gives the required contradiction, which shows us that the real numbers cannot be arranged in a list. We express this idea by saying that the set of real numbers is uncountable.

The set of all real numbers contains so many elements that it is impossible to arrange them in a list. This contrasts with the set $\mathbb{Q}$ of all rational numbers which can be listed. This means that as far as the number of elements is concerned the set of irrational numbers swamps the set of rational numbers. The 'size' of these sets of numbers has been described in a pictorial manner, which is adequate to give a preliminary idea.

A modern computer uses a decimal with a finite number of digits (which must represent a rational number) as an approximation to $\sqrt{2}$ in calculations. We do not question the validity of this as we take it for granted that we can approximate to $\sqrt{2}$ by a terminating decimal to any required degree of accuracy by taking a sufficient number of places after the decimal point. Does the theory vindicate such an assumption? The answer is yes. For we can prove that given any two distinct real numbers $a, b$ with $a < b$, there is a rational number $p/q$ such at $a < p/q < b$ and there is an irrational number $x$ such that $a < x < b$. In particular, for each $n \in \mathbb{Z}^+$ there is a rational number $p/q$ such that $\sqrt{2} - 1/n < p/q < \sqrt{2}$ and so there is a rational number within a distance $1/n$ of $\sqrt{2}$. Similarly, we can find rational numbers within a distance $1/n$ of $\pi$, e, etc. Such statements necessarily require proof.

THEOREM 1.4.6 Let $a, b$ be any two real numbers such that $a < b$. Then there is a rational number $p/q$ and an irrational number $x$ such that $a < p/q < b$ and $a < x < b$.

(A brief geometric description of the problem helps to outline the method of proof and shows how the integers $p, q$ are chosen. The numbers $a, b$ can be represented by points on a line a distance $(b - a)$ apart; see Fig. 1.2.

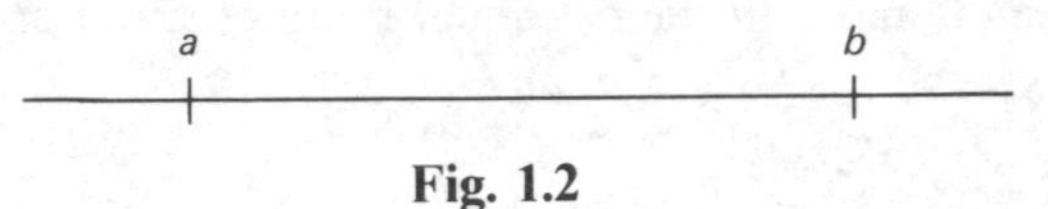


**Fig. 1.2**

In order to ensure that a point representing a number of the form $p/q$ falls between $a$ and $b$ we must choose $q$ sufficiently large so that $1/q < b - a$; i.e. we want $q > 1/(b - a)$. Then we need the smallest integer $p$ such that $p/q > a$. This will give a rational number $p/q$ with $a < p/q < b$. A slight modification of the argument then provides a proof of the existence of an irrational number $x$ with $a < x < b$. Now that the geometrical picture has pin-pointed the salient ideas we can write out a precise proof.)

*Proof* Since $a < b$, we have $b - a > 0$ and $1/(b - a) > 0$. By the Archimedean property there is an integer $q$ such that

$$q > \frac{1}{b - a}.$$

This integer $q$ is positive, since $1/(b - a) > 0$. Moreover,

$$\frac{1}{q} < b - a. \tag{1}$$

Again using the Archimedean property, there is an integer $p'$ such that $p' > qa$. The set of all such integers $p'$ is non-empty and bounded below by $qa$. Using Theorem 1.2.1 and the fact that there is an integer less than $qa$, we see that this set of integers $p'$ has a least element. Let $p$ be this least element. Then

$$p > qa \tag{2}$$

and
$$p - 1 \leqslant qa,$$

i.e.
$$\frac{p}{q} \leqslant \frac{1}{q} + a < (b - a) + a = b \tag{3}$$

using relation (1). Relations (2) and (3) together give

$$a < \frac{p}{q} < b$$

and there are integers $p, q$ such that $a < p/q < b$. Thus there is a rational number $p/q$ between $a$ and $b$.

The proof of the existence of an irrational number $x$ such that $a < x < b$, is simplified if we stipulate that $x$ should be of some definite form. Since $\sqrt{2}$ has received considerable prominence in this chapter, we will seek an irrational number $x$ of the form $p^*\sqrt{2}/q^*$, where $p^*$ and $q^*$ are integers and $q^* > 0$. Obviously the reader could easily adapt the proof to obtain an irrational number of a different form if so desired.

Using the first part, we see that there is a rational number $p^*/q^*$ such that

$$\frac{a}{\sqrt{2}} < \frac{p^*}{q^*} < \frac{b}{\sqrt{2}},$$

i.e.
$$a < \frac{p^*\sqrt{2}}{q^*} < b.$$

Now that it has been proved that there is an irrational $x$ such that $a < x < b$, a further application proves that there are irrationals $x_1, x_2$, such that $a < x_1 < x$ and $x < x_2 < b$. By repeated applications there are an infinite number of irrationals between $a$ and $b$; similarly, there are an infinite number of rationals between $a$ and $b$.

COROLLARY Let $a, b$ be two real numbers such that $a < b$. Then there are an infinite number of rationals $p/q$ with $a < p/q < b$ and an infinite number of irrationals $x$ such that $a < x < b$.

The section closes with a result which demonstrates that the rationals do not satisfy the completeness axiom. Let $S$ be the set of all positive rational numbers $x$ such that $x^2 < 2$,

i.e.
$$S = \{x \in \mathbb{Q} : x > 0 \text{ and } x^2 < 2\}.$$

Then $S$ is not empty. Moreover, if $x \geqslant 2$, then $x^2 \geqslant 4$ and $x \notin S$, the number 2 is, therefore, an upper bound for $S$. Let $A$ be the set of all rational numbers $\alpha$ such that $\alpha$ is an upper bound of $S$,

i.e.
$$A = \{\alpha \in \mathbb{Q} : \alpha \text{ is an upper bound of } S\}.$$

Then it can be proved that $A$ does not have a least member, i.e. the completeness axiom is not satisfied for $\mathbb{Q}$. We first note that if $\alpha \in A$ then $\alpha > \sqrt{2}$. For if $\alpha < \sqrt{2}$, then there is a positive rational number $p/q$ such that $\alpha < p/q < \sqrt{2}$. Since $p/q < \sqrt{2}$, it follows that $(p/q)^2 < 2$ and $p/q \in S$ and so $\alpha$ is not an upper bound of $S$. Thus if $\alpha < \sqrt{2}$ then $\alpha \notin A$. Hence if $\alpha \in A$, then $\alpha \not< \sqrt{2}$. Since $\sqrt{2}$ is irrational and all the elements of $A$ are rational numbers we see that if $\alpha \in A$, then $\alpha > \sqrt{2}$.

Now suppose $\alpha_1$ is any element of $A$. Then $\alpha_1 > \sqrt{2}$ and so there is a rational number $p'/q'$ such that $\alpha_1 > p'/q' > \sqrt{2}$. Moreover if $x \in S$, then $x^2 < 2 < (p'/q')^2$ and so $p'/q'$ is an upper bound of $S$. Hence $p'/q' \in A$ and it follows that if $\alpha_1$ is any element of $A$ then there is an element $p'/q'$ of $A$ with $p'/q' < \alpha_1$. Thus $A$ does not have a least element.

This demonstrates that $\mathbb{Q}$ does not satisfy the completeness axiom, and therefore the completeness axiom certainly distinguishes between $\mathbb{R}$ and $\mathbb{Q}$.

## APPENDIX: MATHEMATICAL INDUCTION

Suppose that there is a statement $T_n$ (involving $n$) associated with each positive integer $n$. For example, $T_n$ might be the statement

$$1 + 2 + 3 + \ldots + n = \tfrac{1}{2}n(n+1) \qquad (n \in \mathbb{Z}^+).$$

It is certainly easy to check that this result is true for the first few cases $n = 1, 2, 3, 4, 5$. To deal with the statement for all $n \in \mathbb{Z}^+$, the principle of mathematical induction is needed.

### Principle of mathematical induction

Suppose that there is a statement $T_n$ associated with each positive integer $n$. If

(a) **the result is true when $n = 1$ (i.e. $T_1$ is true)**, and
(b) **the result is true for $n = k$ whenever it is true for $n = k - 1$** (i.e. $T_{k-1}$ true implies $T_k$ true),

then $T_n$ is true for all positive integer $n$.

It is not difficult to see how this works. The underlying idea is quite simple. Suppose we check in turn the statements for $n = 1$, followed by $n = 2$, then $n = 3$, etc., and we record for each either $T$ for true or $F$ for false. Assumption (a), that the result is true for $n = 1$, means that we begin with $T$. The hypothesis that the result is true for $n = k$ whenever it is true for $n = k - 1$ means that we can never go from $T$ to $F$, i.e. we cannot have a true statement succeeded by a false one. Since the first statement is true, we will record $T, T, T, T, T, T, \ldots$ along the line, and the result is true for all natural numbers. This describes the main features. A proper proof depends on properties of the integers proved in the preceding chapter.

## Proof of the principle of mathematical induction

Suppose the conditions (a) and (b) are satisfied. Then either (I) the result is true for all $n \in \mathbb{Z}^+$ or (II) the result is not true for all $n \in \mathbb{Z}^+$. We will show that (II) leads to a contradiction.

*Case (II)* If the result is not true for all $n \in \mathbb{Z}^+$ then by the corollary to Theorem 1.2.1 there is a smallest natural number $m$ for which the theorem is false and $m \geqslant 2$ because of condition (a). Hence the result is true for $n = m - 1$.

Condition (b) now ensures that the result is true for $n = (m - 1) + 1$, i.e. $n = m$ and we have the required contradiction. Hence case (II) is impossible and so we must have case (I), i.e. $T_n$ is true for all $n \in \mathbb{Z}^+$.

---

### *Examples*

1. Prove that

$$1 + 2 + 3 + \ldots + n = \tfrac{1}{2}n(n+1) \qquad (n \in \mathbb{Z}^+).$$

*Solution* In this case use $T_n$ for the statement

$$1 + 2 + 3 + \ldots + n = \tfrac{1}{2}n(n+1).$$

Then clearly $T_1$ is true. Moreover, if $T_{k-1}$ is true for some $k \geqslant 2$, then

$$1 + 2 + \ldots + (k-1) = \tfrac{1}{2}(k-1)k.$$

Add $k$ to both sides:

$$1 + 2 + \ldots + (k-1) + k = \tfrac{1}{2}(k-1)k + k = \tfrac{1}{2}k(k+1).$$

Hence $T_{k-1}$ true implies $T_k$ is also true. Since $T_1$ is true it follows by mathematical induction that $T_n$ is true for all $n \in \mathbb{Z}^+$.

2. Prove that $5^{3n-1} + 19.3^{4(n-1)}$ is divisible by 44.

*Solution* Let $T_n$ stand for the statement

$$5^{3n-1} + 19.3^{4(n-1)} \text{ is divisible by } 44.$$

When $n = 1$, $5^{3n-1} + 19.3^{4(n-1)} = 5^2 + 19.3^0 = 25 + 19 = 44$ and the result is true when $n = 1$.

Suppose now that the result is true for $n = k - 1$ where $k$ is some integer with $k \geqslant 2$.

i.e.
$$5^{3(k-1)-1} + 19.3^{4(k-1-1)} = 44p,$$

where $p$ is an integer. This gives

$$5^{3k-4} + 19.3^{4(k-2)} = 44p. \qquad (1)$$

Multiply equation (1) by $5^3$ to obtain

$$5^{3k-1} + 19.3^{4(k-2)} \cdot 125 = 125.44p,$$

i.e.
$$5^{3k-1} + 19.3^{4(k-2)} \cdot (81 + 44) = 125.44p,$$

i.e.
$$5^{3k-1} + 19.3^{4(k-1)} + 19.3^{4(k-2)} \cdot 44 = 125.44p,$$

which gives
$$5^{3k-1} + 19.3^{4(k-1)} = 44(125p - 19.3^{4(k-2)}).$$

Thus $5^{3k-1} + 19.3^{4(k-1)}$ is divisible by 44 and it therefore follows that $T_{k-1}$ true implies $T_k$ is also true. Since the result is true when $n = 1$, it follows by the principle of mathematical induction that $T_n$ is true for all $n \in \mathbb{Z}^+$.

*Note:* It is not necessary to start with $n = 1$ as the first case. Thus condition (a) could be replaced by: $T_{n_0}$ is true for some positive integer $n_0$. The conclusion would then be altered to read: $T_n$ is true for all $n \geqslant n_0$.

## *MISCELLANEOUS EXERCISES 1*

**1** The set $A$ consists of all the real numbers of the form $1/2^m + 1/3^n + 1/5^q$ and the set $B$ consists of all the real numbers of the form $1/2^m - 1/3^n$, where $m, n, q$ are positive integers. Decide which of the following exist:

(a) $\sup A$, (b) $\sup B$, (c) $\inf A$, (d) $\inf B$.

Find those which exist.

**2** Let $A$ be a set of positive real numbers with infimum $\alpha > 0$. Let the set $A^{-1}$ be defined by

$$A^{-1} = \left\{ \frac{1}{a} : a \in A \right\}.$$

Show that $A^{-1}$ has a supremum. What is its value?

**3** Decide which of the following statements is true and which is false.

(a) A set of real numbers which has a maximal and minimal element is finite.

(b) Every non-empty set of rational numbers which has a rational upper bound always has a rational supremum.

(c) If $A, B$ are two non-empty sets of real numbers with $\alpha = \sup A$ and $\beta = \sup B$ and if the set $2A + 3B$ is given by

$$2A + 3B = \{2a + 3b : a \in A, b \in B\},$$

then $2A + 3B$ is bounded above and

$$\sup(2A + 3B) = 2\alpha + 3\beta.$$

In the case of a true statement give a proof. In the case of a false statement give a counter-example.

**4** Let $D, E$ be two non-empty sets such that $D \subseteq E$ and $E$ is bounded above. Show that

$$\sup D \leqslant \sup E.$$

**5** Determine the supremum of each of the following sets.

(a) $\left\{1+\frac{1}{n}: n=1,2,3,\ldots\right\}$, (b) $\left\{1-\frac{1}{n}: n=1,2,3,\ldots\right\}$,

(c) $\{x\in\mathbb{R}: 3x^2-10x+4<1\}$.

**6** Let $X$ be a set of positive real numbers with supremum $\alpha$. Write $Y=\{x^2: x\in X\}$. Show that $\alpha^2$ is the supremum of $Y$.

## HINTS FOR SOLUTION OF EXERCISES

### Exercises 1.4.1

**1** (f) The set $S$ is equal to $\{x\in\mathbb{R}: \frac{1}{3}<x<3\}$, since $3x^2-10x+3=(3x-1)(x-3)<0$ if $\frac{1}{3}<x<3$.

### Exercise 1.4.2

**1** Show that the square of a rational number is rational.

**2** Use the fact that the difference of two rational numbers is rational. It follows that if $(x+y)$ is rational and $y$ is rational then $(x+y)-y$ is also rational.

### Miscellaneous Exercises 1

**5** (c) The set is the set of all real numbers $x$ for which $3x^2-10x+3<0$, which we met earlier.

## ANSWERS TO EXERCISES

### Exercises 1.4.1

**1** (a) (i) $\sup S=b$, (ii) $\inf S=a$;
the supremum and the infimum both belong to $S$.

(b) (i) $\sup S=b$; (ii) $\inf S=a$;
neither the supremum nor the infimum belongs to $S$.

(c) (i) $\sup S=\frac{3}{2}$, (ii) $\inf S=0$;
the supremum and infimum both belong to $S$.

(d) (i) $S$ is not bounded above and so there is no supremum;
(ii) $\inf S=0$ and $\inf S$ belongs to $S$.

(e) $S$ is not bounded above and it is not bounded below. Hence $S$ does not possess a supremum and it does not possess an infimum.

(f) (i) $\sup S=3$, (ii) $\inf S=\frac{1}{3}$;
neither the supremum nor the infimum belongs to $S$.

**2** The following are just two possible examples:

(a) $A = B = \{x \in \mathbb{R} : x < 0\}$. In this case $AB = \{x \in \mathbb{R} : x > 0\}$ which is not bounded above.

(b) $A = \{-2, -1, 0\}$, $B = \{-1, 0, 1\}$. In this case $AB = \{-2, -1, 0, 1, 2\}$ and $\sup A = 0$, $\sup B = 1$, $\sup AB = 2$.

## Exercises 1.4.2

**1** The converse is not true. For example, $\sqrt{2}$ is irrational but its square is rational.

**3** False. Counter-example: $x = \sqrt{2}$, $y = 1 - \sqrt{2}$. In this case $x + y$ is rational, but $x$ and $y$ are both irrational.

Not necessarily. If $x = \sqrt{2}$, $y = \sqrt{2}$, then $x$ and $y$ are both irrational, but $xy = 2$, which is rational.

## Miscellaneous Exercises 1

**1** (a) $\sup A = \frac{31}{30}$, (b) $\sup B = \frac{1}{2}$, (c) $\inf A = 0$, (d) $\inf B = -\frac{1}{3}$.

**2** $\sup A^{-1} = 1/\alpha$.

**3** (a) False. Counter-example: the set $S = \{x \in \mathbb{R} : 0 \leqslant x \leqslant 10\}$ has a maximal element 10 and a minimal element 0, but $S$ has an infinite number of elements.

(b) False. Counter-example: let $S = \{x \in \mathbb{Q} : x > 0 \text{ and } x^2 < 2\}$. Then 2 is an upper bound of $S$, for every $x \in S$ satisfies the inequality $x \leqslant 2$. However the supremum of $S$ is $\sqrt{2}$ which is irrational.

(c) True.

**5** (a) The supremum is 2. (b) The supremum is 1.

(c) The supremum is 3.

# 2 SEQUENCES

## 2.1 INTRODUCTION

Early records of classical Greek mathematics show a marked aversion for any methods involving infinite processes. For centuries this remained the prevailing attitude among mathematicians. Indeed, there was no significant change until the time of Sir Isaac Newton. In 1669 he composed his famous treatise *De Analysi per aequationes numero terminorum infinitas* (published in 1711) which included an account of his work on infinite series as well as the beginnings of calculus. Because of the work of Newton the use of infinite processes became regarded as a legitimate mathematical tool. In time, it was recognised that careful rules needed to be developed governing the use of such infinite processes and the calculus in order to ensure the validity of the final results, and thus real analysis came into being. The rigorous subject as we know it today stems, in the main, from the work of eighteenth and nineteenth-century mathematicians on the continent. Many years of work by some of the most outstanding mathematicians of the last few centuries have produced the final polished version.

A fundamental concept is that of a limit and it is probably simplest to introduce it initially in the context of sequences. Certainly an understanding of limits of sequences is needed before we can tackle the problem of manipulating infinite series. Mathematicians of Newton's time were not so careful how they handled infinite series, but then they were occasionally forced to dismiss certain results because they were patently absurd! Had they been in possession of the knowledge of analysis which we have today, they would never have obtained their occasional absurd results, but without them mathematicians might never have seen the need to investigate the problem of what rules are needed to govern the use of infinite processes. Nowadays we have precise laws governing the manipulating of infinite sequences, infinite series and function theory and we can be certain that our results are right provided that we have followed the dictates of analysis correctly. With a view to presenting a logical progression we begin with this chapter on infinite sequences.

## 2.2 SEQUENCES

A sequence of real numbers is a succession of real numbers in a definite order so that we know which number is in the first place, which number is in the second place and for any positive natural number $n$ we know which number is in the $n$th place. Normally, the subscript $n$ is used for the number in the $n$th position and this number is called the $n$th term of the sequence. For example, if the letter $a$ is used for a particular sequence then $a_n$ is normally used for the number in the $n$th position. The sequence then starts $a_1, a_2, a_3, a_4, a_5, a_6, \ldots$. This sequence would be denoted by either $(a_n)_{n=1}^{\infty}$ or just $(a_n)$—the latter notation being used when it is felt that there is no need to emphasise the values which $n$ may assume. The former is particularly useful if the context does not make it clear what values are assumed by $n$ or if we wish to omit a few terms of a sequence. Suppose we take the sequence $a_1, a_2, a_3, a_4, a_5, \ldots$ and we decide to omit the first four terms and take as our new sequence $a_5, a_6, a_7, \ldots$. Then we could write $(a_n)_{n=5}^{\infty}$ for this new sequence and the subscript $n = 5$ at the bottom of the bracket tells us we are going to begin by using $a_5$ as our first term. Of course, many other letters like $b, s, x$, etc. are frequently used for sequences, giving the sequences $(b_n)$, $(s_n)$, $(x_n)$, etc.

---

*Examples 2.2.1*

1. Let $a_n = \frac{1}{n} (n \in \mathbb{Z}^+)$. Then $(a_n)_{n=1}^{\infty}$ is the sequence

$$1, \tfrac{1}{2}, \tfrac{1}{3}, \tfrac{1}{4}, \tfrac{1}{5}, \tfrac{1}{6}, \ldots.$$

2. The sequence $((-1)^n \frac{1}{n})_{n=1}^{\infty}$ starts

$$-1, \tfrac{1}{2}, -\tfrac{1}{3}, \tfrac{1}{4} - \tfrac{1}{5}, \tfrac{1}{6}, -\tfrac{1}{7}, \tfrac{1}{8}, \ldots.$$

3. Let $x_n = \cos n\pi \, (n \in \mathbb{Z}^+)$. Then $(x_n)_{n=1}^{\infty}$ is the sequence

$$-1, +1, -1, +1, -1, +1, \ldots.$$

4. The sequence $(1/n^2)_{n=1}^{\infty}$ starts $1, \frac{1}{4}, \frac{1}{9}, \frac{1}{16}, \ldots$. If we decide to omit the first four terms and start with $\frac{1}{25}, \frac{1}{36}, \frac{1}{49}, \ldots$ then this would be denoted by $(1/n^2)_{n=5}^{\infty}$.

---

Now that we have a general idea of what is meant by a sequence, we can look back and see that in each case the sequence provides a list of real numbers in a definite order. We know precisely which real number is in the $n$th position and so, corresponding to each positive integer $n$, there is a definite real number, viz. the real number in the $n$th position. Thus a sequence provides a mapping from the positive integers $\mathbb{Z}^+$ into the reals. In fact this is probably the best way to formulate the definition of a sequence.

DEFINITION 2.2.1 A sequence of real numbers is a function $f: \mathbb{Z}^+ \to \mathbb{R}$.

With function notation the real number $f(n)$ corresponds to the positive integer $n$. In fact, we hardly ever use function notation. The number $f(n)$ is normally replaced by a number of the form $a_n$, using the common subscript notation.

For sequences the important thing is the order in which the terms appear and the general trend of behaviour as we move along the sequence. Let us start with some familiar examples where the pattern of behaviour seems very obvious.

*Examples 2.2.2*

1. Let $a_n = \frac{1}{n}\,(n \in \mathbb{Z}^+)$. This sequence starts $1, \frac{1}{2}, \frac{1}{3}, \frac{1}{4}, \frac{1}{5}, \frac{1}{6}, \frac{1}{7}, \frac{1}{8}, \ldots$. Without too much thought the reader will inevitably say that this sequence tends to zero, and may very well be happy to write $a_n \to 0$ as $n \to \infty$.
2. Let $a_n = n + (-1)^n \frac{1}{n}\,(n \in \mathbb{Z}^+)$. This sequence starts $0, 2\frac{1}{2}, 2\frac{2}{3}, 4\frac{1}{4}, 4\frac{4}{5}, 6\frac{1}{6}, 6\frac{6}{7}, 8\frac{1}{8}, \ldots$ and again the reader will probably conclude that this sequence tends to infinity and might even write $a_n \to \infty$ as $n \to \infty$.
3. The first example could be modified by alternating the signs $+, -$ to give

$$1, -\tfrac{1}{2}, \tfrac{1}{3}, -\tfrac{1}{4}, \tfrac{1}{5}, -\tfrac{1}{6}, \ldots.$$

   No doubt the reader would still claim (and rightly claim) that the sequence tends to zero.
4. Let $a_n = 1 - \frac{1}{n}\,(n \in \mathbb{Z}^+)$. This sequence begins $0, \frac{1}{2}, \frac{2}{3}, \frac{3}{4}, \frac{4}{5}, \frac{5}{6}, \frac{6}{7}, \frac{7}{8}, \ldots$ and it would seem reasonable to say that $a_n$ tends to 1 as $n \to \infty$.
5. Let $a_n = [\frac{n}{2}] + (-1)^n \frac{1}{2}\,(n \in \mathbb{Z}^+)$, where $[x]$ is the integer part of $x$, i.e. it is the greatest integer $p \leqslant x$. The sequence now starts $-\frac{1}{2}, \frac{3}{2}, \frac{1}{2}, \frac{5}{2}, \frac{3}{2}, \frac{7}{2}, \frac{5}{2}, \frac{9}{2}, \ldots$. The 2000th term $a_{2000}$ is $1000 + \frac{1}{2}$, and the 2 000 000th term is $1\,000\,000 + \frac{1}{2}$. Despite the fact that the size of the terms fluctuates, it still seems reasonable to say that $a_n$ tends to infinity as $n$ tends to infinity. If we draw a graph which shows $a_n$ plotted against $n$ then this claim seems almost self-evident (see Fig. 2.1).

Now let us use these examples to pick out the salient facts. First let us try to decide what properties are necessary if a sequence tends to infinity. Obviously it is not enough to ensure that the terms get larger and larger. For instance, Example 4 provides a sequence whose terms get larger and larger as $n$ increases, but we have already decided that it does not tend to infinity. Indeed, it was claimed that this sequence tends to 1. Furthermore, Example 5 shows a sequence in which the terms do not consistently get larger and larger and yet it does tend to infinity. What leads us to this conclusion? If we consider the graph of the sequence in Example 5, we notice that if we draw a horizontal line across

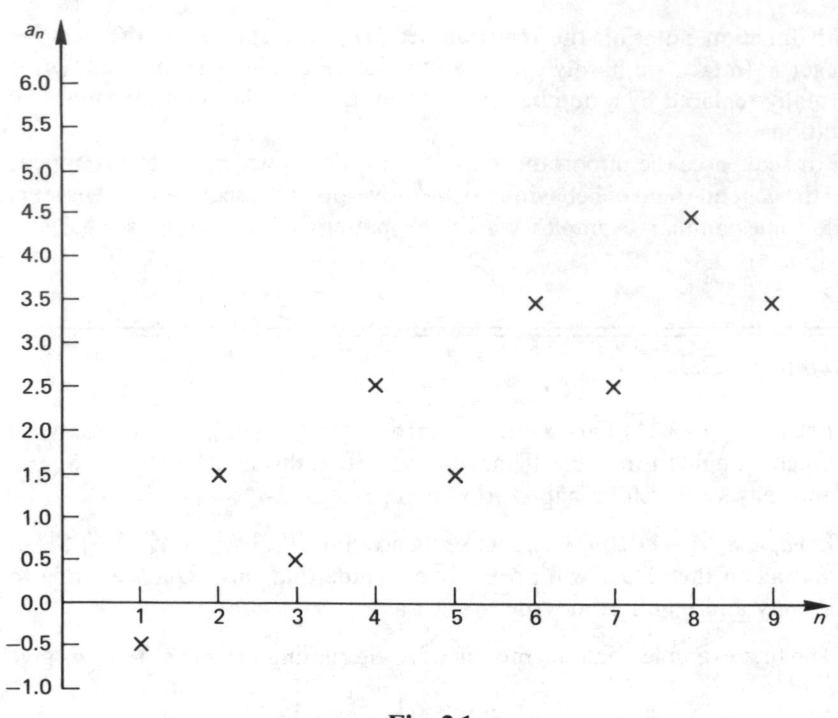


**Fig. 2.1**

the paper at a height 10 above the $n$-axis then all but the first few terms are above this line. Similarly, if we draw a line at a height 100 above the $n$-axis, then all but the first 201 terms are above this line. In general, if we take any positive real number $A$ (however large) and draw a horizontal line across the paper at a height $A$ above the $n$-axis, then all but a finite number of terms at the beginning of the sequence are above the line, i.e. we can find $N$ such that all the terms $a_{N+1}, a_{N+2}, \ldots$ are above the line (see Fig. 2.2). There is therefore some $N$ such that $a_n > A$ for all $n > N$. Obviously, the value taken for $N$ will depend on the size of $A$. For example, if we choose $A = 10$ in Example 5, then we notice $a_n > 10 = A$ for all $n > 21$, i.e. we can use $N = 21$ in this case. For if $n > 21$, then $n \geqslant 22$ and

$$a_n = [\tfrac{n}{2}] + (-1)^n \tfrac{1}{2} \geqslant [\tfrac{22}{2}] + (-1)^n \tfrac{1}{2} = 11 + (-1)^n \tfrac{1}{2} > 10.$$

If we take $A = 100$ in Example 5, then

$$a_n > 100 = A \qquad \text{for all } n > 201.$$

In general, given any $A > 0$, we can always find a corresponding $N$. If we choose $N$ to be the smallest integer such that $N > 2A + 3$, then

$$a_n = [\tfrac{n}{2}] + (-1)^n \tfrac{1}{2} \geqslant [A + 2] + (-1)^n \tfrac{1}{2} > A$$

for all $n > N$, since $n > N$ implies $n \geqslant 1 + N > 2A + 4$.

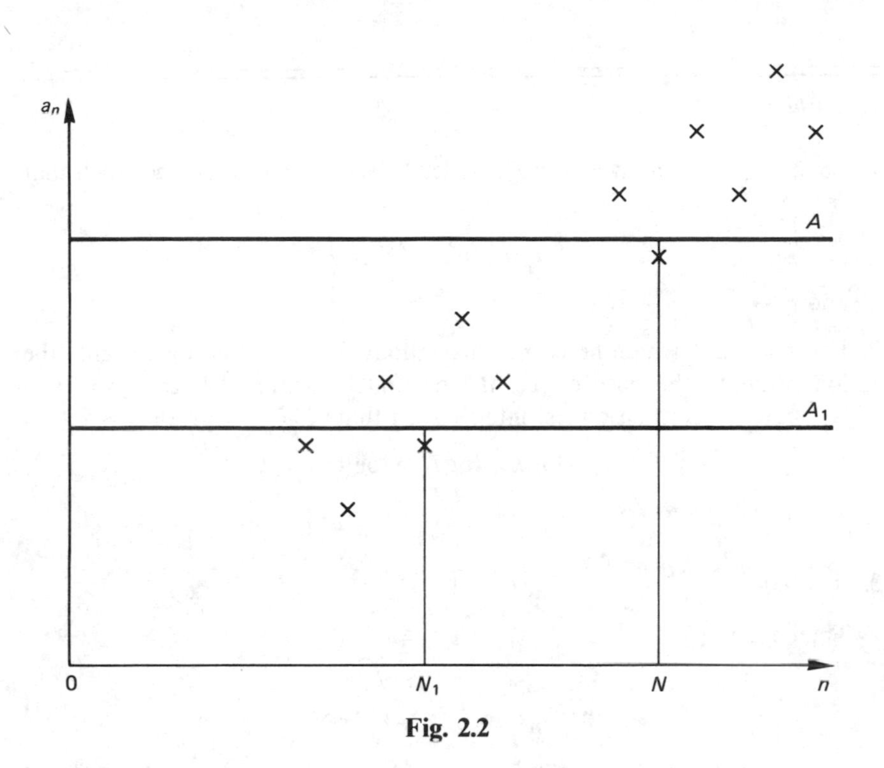


**Fig. 2.2**

Clearly, arguments of the same type could be used for the sequence in Example 2. In this case, given any $A > 0$, all we need to do is to choose $N$ to be the smallest integer greater than $A$. Then, for all $n > N$, $n \geqslant N + 1$ and

$$a_n = n + (-1)^n \tfrac{1}{n} \geqslant N + 1 + (-1)^n \tfrac{1}{n} \geqslant N > A.$$

However, the sequence in Example 4 does not have these properties. For if we choose any number $A$ greater than 1, then all the terms $a_n < A$. We are now in a position to formulate the first definition.

DEFINITION 2.2.2 A sequence $(a_n)$ of real numbers tends to infinity if (and only if), given any $A > 0$ (however large), there exists a corresponding positive integer $N = N(A)$ such that

$$a_n > A \qquad \text{for } \textbf{all } n > N.$$

We write $a_n \to \infty$ as $n \to \infty$.

(In the definition we have written $N = N(A)$ to emphasise the fact that the value chosen for $N$ depends on the value given to $A$ initially. In general, larger values of $A$ give rise to larger values of $N$.)

*Examples 2.2.3*

1. Let $a_n = \sqrt{n}$, then given any $A > 0$, let $N$ be the smallest integer such that $N \geqslant A^2$. For all $n > N$,

$$a_n = \sqrt{n} > \sqrt{N} \geqslant A$$

and $a_n \to \infty$ as $n \to \infty$.

2. Let $a_n = \log n$, where here and throughout this volume 'log' means 'the logarithm to the base e' (i.e. the natural logarithm). Then, given any $A > 0$, let $N$ be the smallest integer such that $N \geqslant e^A$. For all $n > N$,

$$a_n = \log n > \log N \geqslant \log(e^A) = A$$

and $a_n \to \infty$ as $n \to \infty$.

3. Let $a_n = \dfrac{n^2\sqrt{n} + n^2 + 1}{n^2 - 43}$ $\quad (n \geqslant 7)$.

Then, for $n \geqslant 7$,

$$a_n > \frac{n^2\sqrt{n}}{n^2 - 43} > \frac{n^2\sqrt{n}}{n^2} = \sqrt{n}. \tag{1}$$

Given any $A > 0$, choose $N$ to be the smallest integer such that $N \geqslant A^2$ and $N \geqslant 7$, i.e. $N \geqslant \max\{A^2, 7\}$. Then, for all $n > N$,

$$a_n > \sqrt{n} > \sqrt{N} \geqslant A$$

and $a_n \to \infty$ as $n \to \infty$.

In Examples 1 and 2 it was easy to find a simple formula for $N$ in terms of $A$. However, in Example 3 it proved expedient to first use an inequality to obtain $a_n > \sqrt{n}$ and then proceed to obtain a value for $N$ corresponding to any given $A > 0$. For any particular value of $A$, this would not necessarily give the smallest possible corresponding value of $N$, but it will give a value of $N$ which will do. To demonstrate that a value of $N$ exists, we need only find a value which will do; we do not have to find the smallest one possible. Thus in Example 3 we just need to show that there is some $N$ such that

$$\frac{n^2\sqrt{n} + n^2 + 1}{n^2 - 43} > A \qquad \text{for all } n > N;$$

we do not need to find the smallest $N$ satisfying this inequality.

If we write $b_n = \sqrt{n}$, then Example 3 reads as follows: For all $n \geqslant 7$, $a_n > b_n$ and $b_n \to \infty$ as $n \to \infty$. Hence $a_n \to \infty$ as $n \to \infty$.

The reasoning behind this type of method is patently obvious if we use a graphical representation. Draw a graph showing the sequences $(a_n)$ and $(b_n)$ on the same diagram. Then for each $n \geqslant 7$, the cross marking $a_n$ is above the cross for the corresponding $b_n$. Now, if any line is drawn across the paper at a height $A > 0$ above the $n$-axis then there is some $N$ such that $b_n$ is above this line whenever $n > N$. Hence $a_n$ must also be above this line for all $n > N$. In practice, this type of method is extremely useful and it is, therefore, incorporated in a theorem.

THEOREM 2.2.1 Let $(a_n)$ and $(b_n)$ be two sequences of real numbers, such that

$$a_n \geqslant b_n$$

for all $n \geqslant N_0$ (where $N_0$ is some given positive integer). If $b_n \to \infty$ as $n \to \infty$, then this implies that

$$a_n \to \infty \qquad \text{as} \qquad n \to \infty.$$

*Proof* Let $A > 0$. Since $b_n \to \infty$ as $n \to \infty$, there is a positive integer $N'$ such that

$$b_n > A \qquad \text{for all } n > N'.$$

Choose $N = \max\{N_0, N'\}$. Then for all $n > N$,

$$a_n \geqslant b_n > A.$$

Hence $a_n \to \infty$ as $n \to \infty$.

This result is frequently used in conjunction with the triangle inequalities (see appendix at the end of this chapter):

$$||x| - |y|| \leqslant |x + y| \leqslant |x| + |y|;$$
$$||x| - |y|| \leqslant |x - y| \leqslant |x| + |y|.$$

---

*Examples 2.2.4*

1. Let $a_n = n^2 + n \cos n\pi$. Then

$$a_n = n^2 + (-1)^n n \geqslant n^2 - n \geqslant n^2 - \tfrac{1}{2}n^2 = \tfrac{1}{2}n^2 \qquad (n \geqslant 2).$$

Write $b_n = \frac{1}{2}n^2$. Then $b_n \to \infty$ and hence $a_n \to \infty$ as $n \to \infty$.

2. Let $a_n = \dfrac{n^2 + \sqrt{n}}{n + \cos n}$. Then, for $n > 1$,

$$a_n > \frac{n^2}{n + \cos n} \geqslant \frac{n^2}{n+1} > \frac{n^2}{n+n} = \tfrac{1}{2}n.$$

Write $b_n = \frac{1}{2}n$. Since $a_n > b_n (n > 1)$ and $b_n \to \infty$ as $n \to \infty$, we see that $a_n \to \infty$ as $n \to \infty$.

---

## *EXERCISES 2.2.1*

**1** Let $a_n = \log(\log n)$ $(n \geqslant 2)$. Given $A > 0$, find a formula for $N$ such that $a_n > A$ for all $n > N$.

**2** Prove that each of the following sequences $(a_n)$ has the property that $a_n \to \infty$ as $n \to \infty$:

(a) $a_n = n + \cos n$ $(n \in \mathbb{Z}^+)$, (b) $a_n = \dfrac{n^4 + n^3}{3n^3 + 1}$ $(n \in \mathbb{Z}^+)$,

(c) $a_n = \dfrac{n+7}{2 + \sin n}$ $(n \in \mathbb{Z}^+)$.

**3** Show that if $a_n \to \infty$ as $n \to \infty$ and $c$ is a positive real number, then $ca_n \to \infty$ as $n \to \infty$.

The preceding concepts all have analogues corresponding to the case $a_n \to -\infty$ as $n \to \infty$. In this case we need to draw lines across the graph an arbitrary distance $A$ below the $n$-axis, and we require the existence of a corresponding $N$ such that the points representing $a_n$ are below this line for $n > N$ (i.e. $a_n < -A$ for all $n > N$). This gives the following definition.

DEFINITION 2.2.3 Let $(a_n)$ be a sequence of real numbers. Then $a_n \to -\infty$ as $n \to \infty$ if (and only if) given any $A > 0$, there exists a corresponding positive integer $N = N(A)$ such that

$$a_n < -A \qquad \text{for } \textbf{all } n > N.$$

A comparison of the definitions shows immediately that $a_n \to -\infty$ as $n \to \infty$ if and only if $-a_n \to \infty$ as $n \to \infty$. It also suggests that Theorem 2.2.1 has an obvious analogue.

THEOREM 2.2.2 Let $(a_n)$, $(b_n)$ be sequences of real numbers such that

$$a_n \leqslant b_n$$

for all $n \geqslant N_0$, where $N_0$ is some given positive integer. If $b_n \to -\infty$ as $n \to \infty$, then this implies that $a_n \to -\infty$ as $n \to \infty$.

*Proof* Let $A > 0$. Since $b_n \to -\infty$ as $n \to \infty$, there is an integer $N'$ such that

$$b_n < -A \qquad \text{for all } n > N'.$$

Let $N = \max\{N_0, N'\}$. Then for all $n > N$

$$a_n \leqslant b_n < -A$$

and so $a_n \to -\infty$ as $n \to \infty$.

At the beginning of this section we met several examples of sequences which seemed to have a limit which is not infinite. For example, let $a_n = 1 - \frac{1}{n} (n \in \mathbb{Z}^+)$. This gives a sequence which begins with $0, \frac{1}{2}, \frac{2}{3}, \frac{3}{4}, \frac{4}{5}, \ldots$ and appears to tend to 1. Its graph is shown as Fig. 2.3.

We notice that we can ensure that $a_n$ is within a given (arbitrarily small) distance of 1 for all sufficiently large $n$. For example, $1 - \frac{1}{2} < a_n < 1 + \frac{1}{2}$ for all $n > 2$, $1 - \frac{1}{100} < a_n < 1 + \frac{1}{100}$ for all $n > 100$. In fact, if we take a pair of lines across the graph, one at an arbitrarily small distance $\varepsilon$ below 1 and the other $\varepsilon$ above 1, then the points representing $a_n$ are between these lines for all sufficiently large $n$, i.e. $1 - \varepsilon < a_n < 1 + \varepsilon$ for all sufficiently large $n$. For given any $\varepsilon > 0$ (however small) we notice that

$$1 - \varepsilon < a_n < 1 + \varepsilon \qquad \text{for all } n > 1/\varepsilon.$$

If we choose $N$ to be the smallest integer such that $N \geqslant 1/\varepsilon$, then

$$1 - \varepsilon < a_n < 1 + \varepsilon \qquad \text{for all } n > N,$$

i.e.

$$|a_n - 1| < \varepsilon \qquad \text{for all } n > N.$$

These ideas are just what we need to define a finite limit.

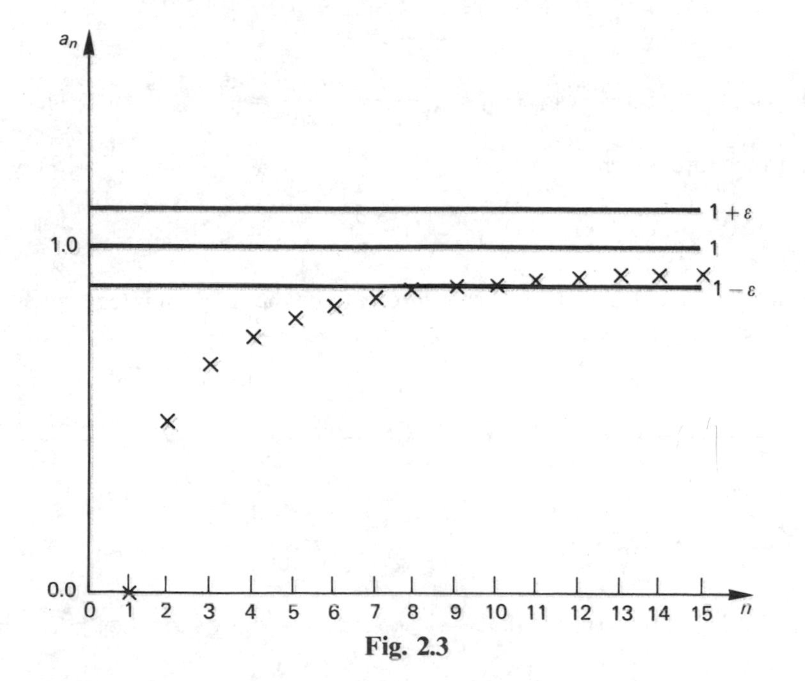


Fig. 2.3

DEFINITION 2.2.4 Let $(a_n)$ be a sequence of real numbers. Then we say that $a_n$ tends to $l$ as $n \to \infty$ if given any $\varepsilon > 0$, there is a corresponding $N = N(\varepsilon)$ such that

$$|a_n - l| < \varepsilon \qquad \text{for } \mathbf{all}\ n > N,$$

i.e.

$$l - \varepsilon < a_n < l + \varepsilon \qquad \text{for } \mathbf{all}\ n > N.$$

We write $a_n \to l$ as $n \to \infty$ or $\lim_{n \to \infty} a_n = l$.

*Note*: The Greek letter $\varepsilon$ has been in common usage in the definition of a limit for over a century. Influenced by Weierstrass's lectures, Heine first used the symbol $\varepsilon$ in his definition of the limit of a function in his *Elemente* in 1872. His definition is in much the same form as we use today.

*Example 2.2.5*

Let $a_n = \dfrac{n^2 + n + 1}{2n^2 + 1}$ $(n \in \mathbb{Z}^+)$. Then it seems reasonable to guess that $a_n \to \frac{1}{2}$ as $n \to \infty$. To justify this assertion we notice that

$$a_n - \frac{1}{2} = \frac{n^2 + n + 1}{2n^2 + 1} - \frac{1}{2}$$

$$= \frac{2n + 1}{2(2n^2 + 1)}.$$

Hence

$$|a_n - \tfrac{1}{2}| < \frac{2n + n}{2 \cdot 2n^2} < \frac{2n + n}{2 \cdot 2n^2} = \frac{3}{4n} \qquad (\text{for } n > 1),$$

and therefore, given $\varepsilon > 0$

$$|a_n - \tfrac{1}{2}| < \frac{3}{4n} < \varepsilon$$

for all $n > \max\{1, \frac{3}{4\varepsilon}\}$. Thus if $N$ is chosen to be the smallest integer such that $N \geqslant \max\{1, \frac{3}{4\varepsilon}\}$, then

$$|a_n - \tfrac{1}{2}| < \varepsilon \qquad \text{for } \mathbf{all}\ n > N.$$

## *EXERCISE 2.2.2*

Use the definition to check that $\dfrac{3n + 1}{n + 2} \to 3$ as $n \to \infty$, by showing that, for any

$\varepsilon > 0$, there is a corresponding $N$ such that

$$\left|\frac{3n+1}{n+2} - 3\right| < \varepsilon \qquad \text{for all } n > N.$$

The preceding work seems to suggest that a sequence cannot have more than one limit and this is indeed true, as we now prove.

THEOREM 2.2.3 A sequence of real numbers cannot have more than one limit.

*Proof* Let $(a_n)$ be a sequence of real numbers. Suppose first that $a_n \to l$ as $n \to \infty$ and $a_n \to m$ as $n \to \infty$. Then either $l = m$ or $l \neq m$. We will show that the assumption $l \neq m$ leads to a contradiction.

*Case $l \neq m$* If $l \neq m$, then $l - m \neq 0$ and $|l - m| > 0$. Using Definition 2.2.3 in the special case $\varepsilon = \frac{1}{2}|l - m|$, we see that as $a_n \to l$, there is an integer $N_0$ such that

$$|a_n - l| < \tfrac{1}{2}|l - m| \qquad \text{for all } n > N_0. \tag{1}$$

Similarly, since $a_n \to m$, there is an integer $N_1$ such that

$$|a_n - m| < \tfrac{1}{2}|l - m| \qquad \text{for all } n > N_1. \tag{2}$$

Now,
$$|l - m| = |l - a_n + a_n - m| \leqslant |l - a_n| + |a_n - m|$$
$$= |a_n - l| + |a_n - m|,$$

and so, for $n > \max\{N_0, N_1\}$, we have

$$|l - m| \leqslant |a_n - l| + |a_n - m| < \tfrac{1}{2}|l - m| + \tfrac{1}{2}|l - m| = |l - m|$$

from (1) and (2). The inequality

$$|l - m| < |l - m|$$

provides the contradiction we were seeking.

Thus the assumption $l \neq m$ leads to a contradiction and is, therefore, untenable. We must therefore have $l = m$. This means that a sequence cannot have two different finite limits.

Moreover, if $a_n \to l$ as $n \to \infty$, then there is $N$ such that

$$l - 1 < a_n < l + 1$$

for all $n > N$. Hence $a_n \nrightarrow \infty$ as $n \to \infty$ and $a_n \nrightarrow -\infty$ as $n \to \infty$. Thus $a_n$ cannot have a finite limit and also an infinite limit.

Finally, $a_n$ cannot tend to both $\infty$ and $-\infty$. The results together show that a sequence cannot have two different limits.

DEFINITION 2.2.5 A sequence $(a_n)$ of real numbers is called a **convergent sequence** if $a_n$ tends to a finite limit as $n \to \infty$.

From Theorem 2.2.3 we see that a convergent sequence has a unique limit.

DEFINITION 2.2.6 If $a_n \to \infty$ as $n \to \infty$ then we say that $(a_n)$ **diverges to infinity**. If $a_n \to -\infty$ as $n \to \infty$, we say that $(a_n)$ diverges to **minus infinity**.

There are, of course, sequences which do not have any limit at all. For example, the sequences $[(-1)^n]_{n=1}^{\infty}$ and $[(-1)^n n]_{n=1}^{\infty}$ do not have any limit.

It would be very difficult if we had to use the definitions every time we needed the limit of a sequence. Using $\varepsilon$ and $N$ each time would involve a tremendous amount of work. It would be rather like going back to first principles in calculus every time a derivative is required. We must, therefore, now develop some new tools for use with sequences.

DEFINITION 2.2.7 A sequence $(a_n)$ of real numbers is called **a bounded sequence** if $|a_n| \leqslant M$ for all $n$ and some positive real number $M$.

THEOREM 2.2.4 A convergent sequence of real numbers is bounded.

*Proof* Let $(a_n)$ be a convergent sequence of real numbers. Then there is some real number $a$ such that $a_n \to a$ as $n \to \infty$. Since $a_n \to a$ as $n \to \infty$, there is a positive integer $N$ such that

$$|a_n - a| < 1 \qquad \text{for all } n > N.$$

Thus $|a_n| < 1 + |a|$ for $n > N$. Let

$$M = \max\{|a_1|, |a_2|, |a_3|, \ldots, |a_{N-1}|, |a_N|, 1 + |a|\}.$$

Then $|a_n| \leqslant M$ for all $n$ and $(a_n)$ is a bounded sequence.

THEOREM 2.2.5 (Algebra of Limits)

If $a_n \to a$ and $b_n \to b$ as $n \to \infty$, then, as $n \to \infty$

(a) $a_n + b_n \to a + b$,
(b) $a_n b_n \to ab$,
(c) $\dfrac{a_n}{b_n} \to \dfrac{a}{b}$, provided $b \neq 0$.

*Proof*

(a) Let $\varepsilon > 0$. Since $a_n \to a$ and $b_n \to b$, there is a positive integer $N$ such that

$$a - \varepsilon/2 < a_n < a + \varepsilon/2 \qquad \text{for all } n > N,$$

$$b - \varepsilon/2 < b_n < b + \varepsilon/2 \qquad \text{for all } n > N.$$

By addition,

$$a + b - \varepsilon < a_n + b_n < a + b + \varepsilon \qquad \text{for all } n > N$$

and $a_n + b_n \to a + b$ as $n \to \infty$.

(b) Since $b_n \to b$ as $n \to \infty$, the sequence $(b_n)$ is convergent and so it is bounded by Theorem 2.2.4. Thus

$$|b_n| \leqslant M \tag{1}$$

for all $n$ and some suitable positive real number $M$. Let $\varepsilon > 0$. Then

$$\frac{\varepsilon}{M + |a|} > 0.$$

Since $a_n \to a$ and $b_n \to b$, there is some $N$ such that

$$|a_n - a| < \frac{\varepsilon}{M + |a|} \qquad \text{for all } n > N \tag{2}$$

and

$$|b_n - b| < \frac{\varepsilon}{M + |a|} \qquad \text{for all } n > N. \tag{3}$$

For all $n > N$,

$$\begin{aligned} |a_n b_n - ab| &= |a_n b_n - a b_n + a b_n - ab| \\ &= |(a_n - a) b_n + a(b_n - b)| \\ &\leqslant |a_n - a||b_n| + |a||b_n - b| \\ &< \frac{\varepsilon}{M + |a|} M + |a| \frac{\varepsilon}{M + |a|} = \varepsilon \end{aligned}$$

using (1), (2) and (3). Thus $a_n b_n \to ab$ as $n \to \infty$.

(c) In view of (b), it is sufficient to prove that $\dfrac{1}{b_n} \to \dfrac{1}{b}$ as $n \to \infty$. Since $b \neq 0$, there is some positive integer $N'$ such that

$$|b_n - b| < \tfrac{1}{2}|b| \qquad \text{for all } n > N'.$$

Hence, for all $n > N'$

$$|b_n| = |b_n - b + b| \geqslant |b| - |b_n - b| > |b| - \tfrac{1}{2}|b| = \tfrac{1}{2}|b|,$$

i.e. $$|b_n| > \tfrac{1}{2}|b| \qquad \text{for all } n > N'. \tag{4}$$

Let $\varepsilon > 0$. Since $b_n \to b$ as $n \to \infty$, there is some $N$ (with $N \geqslant N'$) such that

$$|b_n - b| < \frac{\varepsilon |b|^2}{2} \qquad \text{for all } n > N. \tag{5}$$

Hence, for $n > N$,

$$\left|\frac{1}{b_n} - \frac{1}{b}\right| = \frac{|b - b_n|}{|b_n||b|} < \frac{\varepsilon |b|^2}{2} \cdot \frac{2}{|b|} \cdot \frac{1}{|b|} = \varepsilon$$

from relations (4) and (5) and it follows that $\dfrac{1}{b_n} \to \dfrac{1}{b}$ as $n \to \infty$. It now follows from (b) that $\dfrac{a_n}{b_n} = a_n \cdot \dfrac{1}{b_n} \to a \cdot \dfrac{1}{b} = \dfrac{a}{b}$ as $n \to \infty$.

The proof of the algebra of limits may have appeared long and tedious, but its manifold usefulness amply repays us for all the work, as the next example illustrates.

---

*Example 2.2.6*

Let $$a_n = \frac{n^5 + 7n^3 + 5n^2 + 8}{5n^5 + 3n^4 + 27}.$$

Then $$a_n = \frac{1 + \dfrac{7}{n^2} + \dfrac{5}{n^3} + \dfrac{8}{n^5}}{5 + \dfrac{3}{n} + \dfrac{27}{n^5}} \to \frac{1 + 0 + 0 + 0}{5 + 0 + 0} = \frac{1}{5}$$

as $n \to \infty$, by Theorem 2.2.5.

---

Its use can be greatly extended if it is used in conjunction with the so-called 'sandwich rule' and some standard useful limits. The idea of the 'sandwich rule' or 'squeeze rule' is to sandwich the sequence we are considering between two sequences which are known to have the same limit $l$. The middle sequence must then have the same limit $l$.

THEOREM 2.2.6 (Sandwich Rule) Let $(a_n)$, $(b_n)$, $(c_n)$ be three sequences of real numbers such that $a_n \leqslant b_n \leqslant c_n$ for all $n > N_0$, where $N_0$ is some positive integer. If $a_n \to l$ as $n \to \infty$ and $c_n \to l$ as $n \to \infty$, then $b_n \to l$ as $n \to \infty$.

*Proof* Let $\varepsilon > 0$. Since $a_n \to l$ and $c_n \to l$ as $n \to \infty$ there is some $N$ (with $N \geqslant N_0$) such that

$$l - \varepsilon < a_n < l + \varepsilon \qquad \text{for all } n > N$$

and

$$l - \varepsilon < c_n < l + \varepsilon \qquad \text{for all } n > N.$$

Thus, for all $n > N$,

$$l - \varepsilon < a_n \leqslant b_n \leqslant c_n < l + \varepsilon,$$

and therefore $b_n \to l$ as $n \to \infty$.

THEOREM 2.2.7 Let $(c_n)$ be a sequence of real numbers such that $c_n \geqslant 0$ for all $n > N_0$, where $N_0$ is some given positive real number. Suppose also that $c_n \to c$ as $n \to \infty$. Then

$$c \geqslant 0.$$

*Proof* The proof employs a traditional contradiction argument. Let us suppose that

$$c < 0. \tag{1}$$

Use the definition of a limit with the particular value $\varepsilon = \frac{1}{2}|c|$, i.e. $\varepsilon = -c/2$ because $c < 0$. Since $c_n \to c$ as $n \to \infty$, there is some $N$ (with $N \geqslant N_0$) such that

$$|c_n - c| < \varepsilon = -\tfrac{1}{2}c \tag{2}$$

for all $n > N$. Now $c_n \geqslant 0$ for $n \geqslant N$ and $c < 0$, and hence (2) yields

$$c_n - c = |c_n - c| < -\tfrac{1}{2}c$$

for all $n > N$,

i.e.

$$c_n < \tfrac{1}{2}c < 0$$

(by relation (1)) for all $n > N$. This is impossible as $c_n \geqslant 0$ for all $n > N_0$ and we have the required contradiction. Thus the only possibility is that $c \geqslant 0$ and the proof is complete.

COROLLARY Let $(a_n), (b_n)$ be two sequences of real numbers such that $a_n \leqslant b_n$ for all $n > N_0$, where $N_0$ is some given positive integer. Suppose also that $a_n \to a$, $b_n \to b$ as $n \to \infty$. Then

$$a \leqslant b.$$

*Proof* Use the theorem with $c_n = b_n - a_n$, and the result $b - a \geqslant 0$ is an immediate consequence.

This result shows that weak inequalities are preserved by limits. Is the same true for strict inequalities? Surprisingly the answer is no. For suppose $a_n < b_n$ for all $n > N_0$. Then it is certainly true that $a_n$, $b_n$ must satisfy the inequality $a_n \leqslant b_n$ for all $n > N$. Thus if $a_n \to a$ and $b_n \to b$ as $n \to \infty$, Theorem 2.2.6 guarantees that $a \leqslant b$. It is, however, possible to have $a = b$ and we must, therefore, always use the weak inequality $a \leqslant b$ for the limit.

---

*Example 2.2.7*

Let $a_n = 1/n^2$, $b_n = 1/n$ for $n = 1, 2, 3, \ldots$. Then clearly

$$a_n < b_n$$

for all $n > 1$. However, $a_n \to 0$ and $b_n \to 0$ as $n \to \infty$. The two sequences, therefore, have the same limits despite the fact that $a_n < b_n$ for all $n > 1$.

---

THEOREM 2.2.8 Let $(a_n)$ be a sequence of real numbers. If $a_n \to \infty$ as $n \to \infty$, then $1/a_n \to 0$ as $n \to \infty$.

*Proof* Let $\varepsilon > 0$. Then $1/\varepsilon > 0$. Since $a_n \to \infty$ as $n \to \infty$, there is a positive integer $N$ such that

$$a_n > 1/\varepsilon \qquad \text{for all } n > N.$$

Hence, for all $n > N$,

$$0 < \frac{1}{a_n} < \varepsilon$$

and $1/a_n \to 0$ as $n \to \infty$.

We now return to look at some examples.

---

*Examples 2.2.8*

1. Since $-\dfrac{1}{n} \leqslant \dfrac{\cos n}{n} \leqslant \dfrac{1}{n}$, and $\dfrac{1}{n} \to 0$ as $n \to \infty$, and $-\dfrac{1}{n} \to 0$ as $n \to \infty$, it follows from the sandwich rule that

$$\frac{\cos n}{n} \to 0 \text{ as } n \to \infty.$$

2. $$\frac{n^5+n^4\cos n+6}{4n^5+n^3+\cos n}=\frac{1+\dfrac{\cos n}{n}+\dfrac{6}{n^5}}{4+\dfrac{1}{n^2}+\dfrac{\cos n}{n^5}}\to\frac{1+0+0}{4+0+0}$$

$$=\tfrac{1}{4} \text{ as } n\to\infty.$$

3. Let $(a_n)$ be a sequence of non-negative real numbers and let $a \geqslant 0$. Suppose that $a_n^2 \to a^2$ as $n\to\infty$. Prove that $a_n \to a$ as $n\to\infty$.

*Solution* We consider two separate cases, as follows.

(a) *Case $a=0$* In this case $a_n^2 \to 0$ as $n\to\infty$. If $\varepsilon>0$ is any positive number, then $\varepsilon^2$ is also positive and so there is some $N$ such that

$$|a_n^2|<\varepsilon^2$$

for all $n>N$,

i.e. $$|a_n|<\varepsilon$$

for all $n>N$ and $a_n\to 0$ as $n\to\infty$ as required.

(b) *Case $a>0$* Since $a_n \geqslant 0$ and $a>0$,

$$|a_n^2-a^2|=|(a_n-a)(a_n+a)|=|a_n-a|(a_n+a)\geqslant|a_n-a|a. \qquad (1)$$

Now given any $\varepsilon>0$, the number $\varepsilon a$ is also positive and hence there is some $N$ such that

$$|a_n^2-a^2|<\varepsilon a$$

for all $n>N$. From relation (1) we see that

$$|a_n-a|a\leqslant|a_n^2-a^2|<\varepsilon a$$

for all $n>N$,

i.e. $$|a_n-a|<\varepsilon$$

for all $n>N$ and $a_n\to a$ as $n\to\infty$.

---

The tools we have developed obviously can be used to produce slick and easy answers, but they must not be abused. Before the conclusion of any theorem is applied, the reader must check that all the required conditions are satisfied. If they are not satisfied, then the theorem is not applicable. For example, if $b=0$ the reader must never use the rule $a_n/b_n\to a/b$. The correct use of the algebra of limits will never produce quantities like $\frac{0}{0}$, $\infty/\infty$, $0\cdot\infty$, $\infty-\infty$. If the reader is ever faced by such monstrosities as the answer to a problem then it is as a result of incorrect use of the theorems and further work on these lines is a waste of time. The only solution is to go back to the beginning, start again and remember not to break the rules this time!

The scope of our methods is greatly increased if we add certain standard limits to our array of available tools.

## 2.3 STANDARD LIMITS

### 2.3.1 $n^\alpha$

As $n \to \infty$,

$$n^\alpha \to \begin{cases} \infty & \text{if } \alpha > 0, \\ 1 & \text{if } \alpha = 0, \\ 0 & \text{if } \alpha < 0. \end{cases}$$

For if $\alpha > 0$ and $A > 0$, then $n^\alpha > N^\alpha \geqslant A$ for all $n > N$, where $N$ is the smallest integer such that $\alpha \log N \geqslant \log A$;

i.e. $$N \geqslant e^{(\log A)/\alpha}.$$

Hence $n^\alpha \to \infty$ as $n \to \infty$ for $\alpha > 0$.
When $\alpha = 0$, $n^\alpha = 1$ and therefore $n^\alpha \to 1$ as $n \to \infty$.
When $\alpha < 0$, write $\beta = -\alpha$ so that $\beta > 0$. Then

$$n^\alpha = n^{-\beta} = \frac{1}{n^\beta} \to 0$$

as $n \to \infty$, since $n^\beta \to \infty$ as $n \to \infty$.

### 2.3.2 $a^n$

As $n \to \infty$

$$a^n \to \begin{cases} \infty & \text{if } a > 1, \\ 1 & \text{if } a = 1, \\ 0 & \text{if } |a| < 1. \end{cases}$$

and $a^n$ does not tend to any limit if $a \leqslant -1$.

*If $a > 1$,* we can write $a = 1 + h$ where $h > 0$.

Hence $$a^n = (1+h)^n = 1 + nh + \binom{n}{2}h^2 + \ldots + h^n \geqslant 1 + nh.$$

Since $h > 0$, $1 + nh \to \infty$ as $n \to \infty$ and therefore $a^n \to \infty$ as $n \to \infty$ when $a > 1$.

*If $a = 1$,* then $a^n = 1$ for all $n$ and $a^n \to 1$ as $n \to \infty$.

*If $0 < a < 1$,* then we can write $a = 1/b$ where $b > 1$. Thus

$$a^n = \frac{1}{b^n} \to 0 \quad \text{as} \quad n \to \infty,$$

since $b^n \to \infty$ as $n \to \infty$.

*If $a = 0$,* then $a^n = 0$ for all $n$ and $a^n \to 0$ as $n \to \infty$.

*If $-1 < a < 0$,* then

$$-|a|^n \leqslant a^n \leqslant |a|^n$$

and $|a|^n \to 0$ as $n \to \infty$ because $0 < |a| < 1$.

By the sandwich rule $a^n \to 0$ as $n \to \infty$.

*If* $a \leqslant -1$, then $a^n = (-1)^n |a|^n$.

Since $|a|^n \geqslant 1$ for all $n$, $a^n$ cannot tend to any limit as $n \to \infty$.

### 2.3.3 $a^n/n^\alpha$

For $a > 1$ and $\alpha > 0$, it has already been proved that $a^n \to \infty$ and $n^\alpha \to \infty$ as $n \to \infty$, but what happens to the quotient? Does it tend to any limit at all? If it does have a limit, is the limit finite or infinite? Suprisingly, the answer is that $a^n/n^\alpha \to \infty$ as $n \to \infty$ for $a > 1$ and $\alpha > 0$. If we take $a$ close to 1 and $\alpha$ very large, then this seems at first sight somewhat unexpected. For example, we are claiming that

$$\frac{(1.000\,000\,001)^n}{n^{1\,000\,000\,001}} \to \infty$$

as $n \to \infty$, which seems quite incredible and the use of a large modern computer to find the first million terms would be of little help in making this look even plausible. Yet our analysis shows that the claim is indeed true. The proof is not unduly difficult. The first step is to introduce an integral value for the power of $n$. Let $a > 1$, $\alpha > 0$ and choose an integer $p$ such that $p \geqslant \alpha$. Then

$$\frac{a^n}{n^\alpha} \geqslant \frac{a^n}{n^p} \tag{1}$$

and it is sufficient to show that $a^n/n^p \to \infty$ as $n \to \infty$. The actual details of the proof can often be obscured by the notation in the general case and so we illustrate the method by looking initially at the special case $p = 2$. Since $a > 1$, we can write $a = 1 + h$, where $h > 0$. Thus for $n \geqslant 4$

$$\begin{aligned}\frac{a^n}{n^2} &= \frac{(1+h)^n}{n^2} = \frac{1}{n^2}\left(1 + nh + \frac{n(n-1)}{2!}h^2 + \frac{n(n-1)(n-2)}{3!}h^3 + \ldots + h^n\right)\\ &> \frac{n(n-1)}{n^2}\frac{(n-2)}{3!}h^3 = \frac{(n-1)(n-2)}{6n}h^3\\ &> \frac{\frac{1}{2}n \cdot \frac{1}{2}n}{6n}h^3 = \frac{nh^3}{24}.\end{aligned}$$

Now $nh^3/24 \to \infty$ as $n \to \infty$ and, therefore $a^n/n^2 \to \infty$ as $n \to \infty$. A similar argument can now be used to show that for all positive integers $p$, $a^n/n^p \to \infty$ as $n \to \infty$. In view of (1) and Theorem 2.2.1 we now have $a^n/n^\alpha \to \infty$ as $n \to \infty$ for $a > 1$ and $\alpha > 0$.

Frequently it is useful to take the reciprocal of $a^n/n^\alpha$. Since $a^n/n^\alpha \to \infty$ as $n \to \infty$, it follows that $n^\alpha/a^n \to 0$ as $n \to \infty$ for $a > 1$ and $\alpha > 0$. The substitution $b = 1/a$ then gives the useful form of the limit, viz.

$$n^\alpha b^n \to 0 \text{ as } n \to \infty \qquad \text{for } 0 < b < 1 \text{ and } \alpha > 0.$$

By using the sandwich rule this result can be extended to show that for $\alpha > 0$ and $|b| < 1$,

$$n^{\alpha}b^n \to 0 \qquad \text{as } n \to \infty.$$

## *EXERCISE 2.3.1*

Choose any positive real number $a$ and use a pocket calculator to find the first few terms of the sequence $(a^{1/n})_{n=1}^{\infty}$. How does $a^{1/n}$ behave as $n$ increases? Try a few more positive values for $a$ and check what happens to $a^{1/n}$.

### 2.3.4 $a^{1/n}$

No doubt the previous exercise will have led the reader to the conclusion that for all $a > 0$, $a^{1/n} \to 1$ as $n \to \infty$. Is this really correct? The answer is yes, as we now prove.

First assume $a > 1$ and write $a^{1/n} = 1 + h_n$. Since $a^{1/n} > 1$ when $a > 1$, we see that $h_n > 0$. Moreover,

$$\begin{aligned} a = (1 + h_n)^n &= 1 + nh_n + \frac{n(n-1)}{2!}h_n^2 + \ldots + h_n^n \\ &> nh_n, \end{aligned}$$

and therefore

$$0 < h_n < \frac{a}{n}.$$

By the sandwich rule, $h_n \to 0$ as $n \to \infty$ and $a^{1/n} = 1 + h_n \to 1 + 0 = 1$ as $n \to \infty$.
If $a = 1$, then $a^{1/n} = 1$ for all $n$ and $a^{1/n} \to 1$ as $n \to \infty$.
If $0 < a < 1$, then we can write $a = 1/b$, where $b > 1$. Then

$$a^{1/n} = \left(\frac{1}{b}\right)^{1/n} = \frac{1}{b^{1/n}} \to \frac{1}{1} = 1$$

as $n \to \infty$.

Thus for all $a > 0$, $a^{1/n} \to 1$ as $n \to \infty$.

### 2.3.5 $n^{1/n}$

Using a method similar to the one above, it can be shown that $n^{1/n} \to 1$ as $n \to \infty$. Write $n^{1/n} = 1 + k_n$. If $n > 1$, then $n^{1/n} > 1$ and $k_n > 0$ for $n > 1$.

Now, for $n \geqslant 2$,

$$n = (1 + k_n)^n = 1 + nk_n + \frac{n(n-1)}{2!}k_n^2 + \ldots + k_n^n > \frac{n(n-1)}{2!}k_n^2$$

and therefore

$$0 < k_n < \sqrt{\left(\frac{2}{n-1}\right)} \qquad (n \geqslant 2).$$

By the sandwich rule $k_n \to 0$ as $n \to \infty$ and $n^{1/n} = 1 + k_n \to 1 + 0 = 1$ as $n \to \infty$.

### 2.3.6 $a^n/n!$

Let $a$ be any real number. Choose $N$ to be the smallest positive integer such that $N \geqslant 2|a|$. Then for each integer $p \geqslant N$,

$$\frac{|a|}{p} \leqslant \frac{|a|}{N} \leqslant \frac{1}{2}. \tag{1}$$

For all $n > N$,

$$0 \leqslant \left|\frac{a^n}{n!}\right| = \frac{|a|}{n} \cdot \frac{|a|}{(n-1)} \cdots \frac{|a|}{(N+1)} \cdot \frac{|a|^N}{N!} \leqslant \frac{1}{2} \cdot \frac{1}{2} \cdots \frac{1}{2} \frac{|a|^N}{N!}$$

$$= \left(\frac{1}{2}\right)^{n-N} \frac{|a|^N}{N!}$$

and $\left(\frac{1}{2}\right)^{n-N} \to 0$ as $n \to \infty$. Hence $\left|\frac{a^n}{n!}\right| \to 0$ as $n \to \infty$ and so $\frac{a^n}{n!} \to 0$ as $n \to \infty$.

### 2.3.7 $n!/n^n$

For all $n$

$$0 < \frac{n!}{n^n} = \frac{n(n-1)\ldots 2 \cdot 1}{n^n} = 1\left(1 - \frac{1}{n}\right)\left(1 - \frac{2}{n}\right)\ldots\frac{2}{n} \cdot \frac{1}{n} \leqslant \frac{1}{n}.$$

By the sandwich rule $n!/n^n \to 0$ as $n \to \infty$.

By using the algebra of limits, the sandwich rule and the standard limits we can now deal fairly simply with a large variety of sequences.

---

*Examples 2.3.1*

1. $\dfrac{n^7 7^n + n^5 5^n}{3^n + 8^n} = \dfrac{n^7(\frac{7}{8})^n + n^5(\frac{5}{8})^n}{(\frac{3}{8})^n + 1} \to \dfrac{0+0}{0+1} = 0$

   as $n \to \infty$ using the algebra of limits and subsections 2.3.2 and 2.3.3.

2. $\dfrac{n! + 8^n}{7^n + n!} = \dfrac{1 + \dfrac{8^n}{n!}}{\dfrac{7^n}{n!} + 1} \to \dfrac{1+0}{0+1} = 1$

   as $n \to \infty$ using the algebra of limits and subsection 2.3.6.

3. Find the value of $\lim_{n\to\infty}(4^{10}+r^n)^{1/n}$, where $r$ is any positive real number.

*Solution*
If $0<r\leqslant 1$, then

$$(4^{10})^{1/n}\leqslant(4^{10}+r^n)^{1/n}\leqslant(4^{10}+1)^{1/n}.$$

Now as $n\to\infty$, $(4^{10})^{1/n}\to 1$ and $(4^{10}+1)^{1/n}\to 1$.
By the sandwich rule,

$$\lim_{n\to\infty}(4^{10}+r^n)^{1/n}=1$$

when $0<r\leqslant 1$.
If $r>1$, then $r^n\to\infty$ as $n\to\infty$ and so there is an integer $N$ such that $r^n>4^{10}$ for all $n>N$. It follows that for $n>N$

$$r=(r^n)^{1/n}<(4^{10}+r^n)^{1/n}<(r^n+r^n)^{1/n}=2^{1/n}\cdot r.$$

Since $2^{1/n}\to 1$ as $n\to\infty$, the sandwich rule shows that

$$\lim_{n\to\infty}(4^{10}+r^n)^{1/n}=r$$

when $r>1$.

## EXERCISES 2.3.2

Determine whether the following sequences have a limit. If the limit exists then find it.

**1** $\left(\cos\left(\frac{n}{2^n}\right)\right)$.

**2** $\left(\frac{n^2 10^n+n^3 9^n}{1+7^{2n}}\right)$.

**3** $\left(\frac{n+(-1)^n\sqrt{n}}{(n^2+1)^{1/2}}\right)$.

**4** $\left(\sin\left[\frac{n^2+5}{n+5^n}\right]\right)$.

**5** $\left(\frac{1}{\sqrt{n}(\sqrt{n^2+2}-\sqrt{n^2+1})}\right)$.

**6** $\left(\frac{3^n+4^n}{2^n+5^n}\right)$.

**7** $\left(\frac{2^n+5^n}{3^n}\right)$.

**8** $\left(\frac{100^n+2^n}{n!+99^n}\right)$.

**9** $\left(\frac{3n^2+2n+1}{n^2+1}\right)$.

**10** $\left(\frac{3n^2+2n+1}{n+1}\right)$.

**11** $([2^{10}+(\tfrac{1}{2})^n]^{1/n})$.

**12** $([2^{10}+3^n]^{1/n})$.

**13** $\left(\frac{n^3-n^2\cos n+2}{4n^3+n^2-4\sin n}\right)$.

**14** $\left(\left[\frac{n+2}{3n+4}\right]^5\right)$.

Give reasons for your answers.

## 2.4 SOME GENERAL RESULTS FOR SEQUENCES

We begin by considering increasing sequences and decreasing sequences. Not surprisingly, we start in the time-honoured way with some definitions. The first book of Euclid's elements opens abruptly with twenty-three definitions—it has no introduction at all. Initially we content ourselves with a mere five definitions!

DEFINITION 2.4.1 A sequence $(a_n)$ of real numbers is called an **increasing** sequence if $a_n \leqslant a_{n+1}$ for all $n$. It is called a **strictly increasing** sequence if $a_n < a_{n+1}$ for all $n$.

In a way the terminology is very obvious. A sequence whose terms strictly increase as $n$ increases is called a strictly increasing sequence.

DEFINITION 2.4.2 A sequence $(a_n)$ of real numbers is called a **decreasing** sequence if $a_n \geqslant a_{n+1}$ for all $n$. It is called **strictly decreasing** if $a_n > a_{n+1}$ for all $n$.

DEFINITION 2.4.3 A sequence of real numbers is called a **monotonic** sequence if it is either an increasing sequence or a decreasing sequence.

DEFINITION 2.4.4 A sequence $(a_n)$ of real numbers is said to be **bounded above** if the set $\{a_n : n \in \mathbb{Z}^+\}$ is bounded above, i.e. if there is some $M$ such that $a_n \leqslant M$ for all $n$.

DEFINITION 2.4.5 A sequence $(a_n)$ of real numbers is said to be **bounded below** if the set $\{a_n : n \in \mathbb{Z}^+\}$ is bounded below, i.e. if there is some $m$ such that $m \leqslant a_n$ for all $n$.

If we draw graphs illustrating increasing sequences, then it looks as if there are only two possible types of behaviour. Either the sequence is bounded above and tends to a finite limit or the sequence is not bounded above and tends to infinity. Using the completeness axiom it can easily be shown that this is indeed the case.

THEOREM 2.4.1 Let $(a_n)$ be an increasing sequence of real numbers. Then either

(a) $(a_n)$ is bounded above and $a_n$ tends to a finite limit as $n \to \infty$, or
(b) $(a_n)$ is not bounded above and $a_n \to \infty$ as $n \to \infty$.

*Proof*

(a) Let $S=\{a_n:n\in\mathbb{Z}^+\}$. If $(a_n)$ is bounded above, then the set $S$ is non-empty and bounded above. By the completeness axiom it has a supremum. Let $a=\sup S$. Then, given any $\varepsilon>0$, $a-\varepsilon$ is not an upper bound of $S$. Hence there is some $x_N\in S$ such that

$$a-\varepsilon<x_N\leqslant a.$$

Now the sequence $(a_n)$ is increasing, and therefore for all $n>N$,

$$a-\varepsilon<x_N\leqslant x_n\leqslant a.$$

Hence $x_n\to a$ as $n\to\infty$.

(b) If $(a_n)$ is not bounded above, then, given $A>0$, there must be some term $x_N$ such that

$$x_N>A.$$

Since $(x_n)$ is increasing, $\quad x_n\geqslant x_N>A$

for all $n>N$ and so $x_n\to\infty$ as $n\to\infty$.

Since a convergent sequence is bounded (see Theorem 2.2.4), we have the following corollary.

COROLLARY An increasing sequence of real numbers converges if and only if it is bounded above.

*Note:* The reader is advised to look back to Definition 2.2.5 in which a convergent sequence is defined to be one with a **finite** limit.

Similar results hold for decreasing sequences.

THEOREM 2.4.2 Let $(a_n)$ be a decreasing sequence of real numbers. Then either

(a) $(a_n)$ is bounded below and $(a_n)$ converges, or
(b) $(a_n)$ is not bounded below and $a_n\to-\infty$ as $n\to\infty$.

COROLLARY A decreasing sequence of real numbers converges if and only if it is bounded below.

These results are readily proved by applying the previous theorem and its corollary to the sequence $(-a_n)$.

The previous two theorems, taken together, give the following theorem.

THEOREM 2.4.3 A bounded monotonic sequence converges.

We first apply Theorem 2.4.1 to the sequence $[(1+\frac{1}{n})^n]_{n=1}^{\infty}$ to obtain a very useful standard limit. As $n \to \infty$, $1+\frac{1}{n} \to 1$ and it is a very common error to imagine that this implies that $(1+\frac{1}{n})^n$ also tends to the same limit. However, this seems much less likely when one realises that for each $n \in \mathbb{Z}^+$

$$\begin{aligned}\left(1+\frac{1}{n}\right)^n &= 1+n\left(\frac{1}{n}\right)+\frac{n(n-1)}{2!}\left(\frac{1}{n}\right)^2+\dots\\ &= 1+1+\frac{(n-1)}{2n}+\dots\\ &\geqslant 2.\end{aligned}$$

Thus each of the terms $(1+1/n)^n \geqslant 2$, and, if there is a limit, this limit must be at least 2. A quick calculation of the first few terms suggests that the sequence is increasing, which is indeed true. For let

$$a_n = \left(1+\frac{1}{n}\right)^n.$$

Then

$$\begin{aligned}a_n &= 1+n\left(\frac{1}{n}\right)+\frac{n(n-1)}{2!}\left(\frac{1}{n}\right)^2+\dots+\frac{n!}{n!}\left(\frac{1}{n}\right)^n\\ &= 1+1+\frac{1}{2!}\left(1-\frac{1}{n}\right)+\dots+\frac{1}{r!}\left(1-\frac{1}{n}\right)\left(1-\frac{2}{n}\right)\dots\left(1-\frac{r-1}{n}\right)+\dots\\ &\quad+\frac{1}{n!}\left(1-\frac{1}{n}\right)\dots\left(1-\frac{n-1}{n}\right)\end{aligned}$$

$$\begin{aligned}a_{n+1} &= 1+1+\frac{1}{2!}\left(1-\frac{1}{n+1}\right)+\dots\\ &\quad+\frac{1}{r!}\left(1-\frac{1}{n+1}\right)\left(1-\frac{2}{n+1}\right)\dots\left(1-\frac{r-1}{n+1}\right)+\dots\\ &\quad+\frac{1}{n!}\left(1-\frac{1}{n+1}\right)\dots\left(1-\frac{n-1}{n+1}\right)+\frac{1}{(n+1)!}\left(1-\frac{1}{n+1}\right)\dots\left(1-\frac{n}{n+1}\right).\end{aligned}$$

The first two terms of $a_n$ and $a_{n+1}$ are the same; the $r$th term of $a_{n+1}$ is greater than the $r$th term of $a_n$ for $3 \leqslant r \leqslant n+1$. Finally, $a_{n+1}$ has an extra positive term

$$\frac{1}{(n+1)!}\left(1-\frac{1}{n+1}\right)\dots\left(1-\frac{n}{n+1}\right)$$

at the end. Hence $a_n < a_{n+1}$ and $(a_n)$ is a strictly increasing sequence. Moreover,

$$a_n = 1+1+\frac{1}{2!}\left(1-\frac{1}{n}\right)+\frac{1}{3!}\left(1-\frac{1}{n}\right)\left(1-\frac{2}{n}\right)+\dots+\frac{1}{n!}\left(1-\frac{1}{n}\right)\dots\left(1-\frac{n-1}{n}\right)$$

$$\leqslant 1+1+\frac{1}{2!}+\frac{1}{3!}+\ldots+\frac{1}{n!}$$

$$\leqslant 1+1+\frac{1}{2}+\frac{1}{2\cdot 2}+\ldots+\frac{1}{2^{n-1}}$$

since $$\frac{1}{r!}=\frac{1}{r(r-1)\ldots 2\cdot 1}\leqslant\frac{1}{2\cdot 2\cdot 2\ldots 2\cdot 1}=\frac{1}{2^{r-1}} \qquad (r\geqslant 2).$$

Hence $$a_n<1+\frac{1-(\frac{1}{2})^n}{1-\frac{1}{2}}<1+\frac{1}{1-\frac{1}{2}}=1+2=3,$$

and the sequence $(a_n)$ is also bounded above. By Theorem 2.4.1 $(a_n)$ converges, i.e. $a_n$ tends to a finite limit as $n\to\infty$. We denote the value of this limit by $e$, which gives

$$\left(1+\frac{1}{n}\right)^n\to e \qquad \text{as } n\to\infty.$$

*Note:* This is a special case of the rule $(1+\frac{x}{n})^n\to e^x$ as $n\to\infty$, which will be proved later in the volume.

Sometimes sequences are defined by a recurrence relation rather than an explicit formula. Let us look at a particular example.

---

*Example 2.4.1*

The sequence $(a_n)$ is defined by

$$a_1=\tfrac{5}{2},\ 5a_{n+1}=a_n^2+6 \qquad (n=1,2,3,\ldots).$$

(a) Using induction it can be proved that

$$2<a_n<3$$

for all $n$. Obviously the result is true when $n=1$, since $a_1=\frac{5}{2}$ by definition. Now, suppose that it is true for $n=k-1$, where $k$ is some integer such that $k\geqslant 2$. Then

$$2<a_{k-1}<3. \tag{1}$$

Now, $$5a_k=a_{k-1}^2+6<3^2+6=15$$

by relation (1), and also

$$5a_k=a_{k-1}^2+6>2^2+6=10.$$

Hence $2<a_k<3$ and the assumption that $2<a_{k-1}<3$ implies that $2<a_k<3$. Since the result is true when $n=1$, it follows by induction that

$$2 < a_n < 3 \tag{2}$$

for all $n$.

(b) We now show that $(a_n)$ decreases. Using the definition, we see that

$$\begin{aligned} a_{n+1} - a_n &= \tfrac{1}{5}(a_n^2 + 6) - a_n \\ &= \tfrac{1}{5}(a_n^2 - 5a_n + 6) \\ &= \tfrac{1}{5}(a_n - 3)(a_n - 2) < 0 \end{aligned}$$

since $2 < a_n < 3$ from (a). Hence $(a_n)$ is decreasing. Since it is bounded below it converges. Let $a$ be its limit. Then $a_n \to a$ as $n \to \infty$ and so $a_{n+1} \to a$ as $n \to \infty$. But

$$a_{n+1} = \tfrac{1}{5}(a_n^2 + 6) \to \tfrac{1}{5}(a^2 + 6)$$

as $n \to \infty$. Since the limit is unique,

$$a = \tfrac{1}{5}(a^2 + 6),$$

i.e.
$$a^2 - 5a + 6 = 0$$

which gives $a = 2$ or $a = 3$. Since the sequence is decreasing, $a \leqslant a_1 = \frac{5}{2}$ and so the only possibility is $a = 2$.

---

## *EXERCISE 2.4.1*

The sequence $(a_n)$ is defined by $a_1 = 1$, $a_{n+1} = 4 + 2a_n^{1/3}$ $(n \in \mathbb{Z}^+)$. Use mathematical induction to prove that $1 \leqslant a_n \leqslant 8$ for all $n \in \mathbb{Z}^+$. Show also that $(a_{n+1}/a_n) \geqslant 1$ for all $n$. Deduce that $(a_n)$ converges and find its limit.

## 2.5 SUBSEQUENCES

If we take a sequence $(a_n)$ and delete either a finite or an infinite number of terms then the resulting progression is called a subsequence of the original sequence. For example, suppose we delete all except every third term. This gives $a_3, a_6, a_9, a_{12}, \ldots$ and this is a subsequence of the original sequence. Normally the first term of the subsequence is given the subscript $n_1$, the second is given the subscript $n_2$, and so on. Thus we would write $a_{n_1}, a_{n_2}, a_{n_3}, a_{n_4}, \ldots$ for a subsequence of $a_1, a_2, a_3, \ldots$. Normally such a subsequence would be denoted by $(a_{n_k})_{k=1}^{\infty}$. We notice that we must keep to the order of the original sequence and so

$$n_1 < n_2 < n_3 < n_4 < \ldots.$$

DEFINITION 2.5.1 Let $(n_k)_{k=1}^{\infty}$ be a strictly increasing sequence of positive integers, i.e. $1 \leqslant n_1 < n_2 < n_3 < \ldots$. Then $(a_{n_k})_{k=1}^{\infty}$ is called a **subsequence** of $(a_n)_{n=1}^{\infty}$.

Thus a subsequence is obtained by taking the original sequence and deleting terms (without any change of order).

*Example 2.5.1*

Let $a_n = \frac{1}{n}$ $(n \in \mathbb{Z}^+)$. Then the sequence $(a_n)$ begins $1, \frac{1}{2}, \frac{1}{3}, \frac{1}{4}, \frac{1}{5}, \frac{1}{6}, \ldots$. Hence $1, \frac{1}{3}, \frac{1}{5}, \frac{1}{7}, \ldots$ is a subsequence of $(a_n)$. Another example of a subsequence is $\frac{1}{2}, \frac{1}{4}, \frac{1}{6}, \frac{1}{8}, \ldots$. We could of course produce a subsequence by deleting all the terms $\frac{1}{n}$ for which $n$ is not a prime. This would leave the subsequence

$$\frac{1}{2}, \frac{1}{3}, \frac{1}{5}, \frac{1}{7}, \frac{1}{11}, \frac{1}{13}, \frac{1}{17}, \frac{1}{19}, \frac{1}{23}, \ldots.$$

As we might expect, every subsequence of a convergent sequence has the same limit as the original sequence.

THEOREM 2.5.1 Let $(a_n)$ be a convergent sequence of real numbers and suppose $a_n \to a$ as $n \to \infty$. Then every subsequence also converges to $a$.

*Proof* Let $(a_{n_k})_{k=1}^{\infty}$ be any subsequence of $(a_n)$. Then

$$n_1 < n_2 < n_3 < n_4 < \ldots$$

and

$$n_k \geqslant k \tag{1}$$

for all $k$. Since $a_n \to a$ as $n \to \infty$, given any $\varepsilon > 0$ there is a corresponding $N$ such that

$$|a_n - a| < \varepsilon \qquad \text{for all } n > N. \tag{2}$$

Now if $k > N$, then $n_k > N$ by (1) and therefore

$$|a_{n_k} - a| < \varepsilon \qquad \text{for all } k > N.$$

Hence $a_{n_k} \to a$ as $k \to \infty$.

COROLLARY If $(a_n)$ has two subsequences which converge to different limits then $(a_n)$ cannot converge.

Using a similar proof we can show that the following results hold for sequences which diverge to infinity or to minus infinity.

THEOREM 2.5.2 Let $(a_n)$ be a sequence of real numbers such that $a_n \to \infty$ as $n \to \infty$. Then every subsequence of $(a_n)$ also diverges to $\infty$.

THEOREM 2.5.3 Let $(a_n)$ be a sequence of real numbers such that $a_n \to -\infty$ as $n \to \infty$. Then every subsequence of $(a_n)$ also diverges to $-\infty$.

*Examples 2.5.2*

1. Let $a_n = (-1)^n$. Then $a_{2n} = +1$, $a_{2n+1} = -1$. Thus the subsequence $a_2, a_4, a_6, \ldots$ converges to 1 and the subsequence $a_1, a_3, a_5, \ldots$ converges to $-1$. Since there are two different subsequences with two different limits we see that $a_n$ does not tend to any limit as $n \to \infty$.
2. Let $b_n = a^n$ where $a$ is some real number with $a \leqslant -1$.

   Then $$b_{2n} = a^{2n} = |a|^{2n} \text{ and } b_{2n-1} = a^{2n-1} = -|a|^{2n-1}.$$

   *If* $a = -1$, the subsequence $(b_{2n})_{n=1}^{\infty}$ converges to $+1$ and the subsequence $(b_{2n-1})_{n=1}^{\infty}$ converges to $-1$. Hence $a^n$ cannot tend to any limit as $n \to \infty$.

   *If* $a < -1$, the subsequence $(b_{2n})_{n=1}^{\infty}$ diverges to $\infty$ and the subsequence $(b_{2n-1})_{n=1}^{\infty}$ diverges to $-\infty$. Hence, again, $a^n$ does not tend to any limit as $n \to \infty$.
3. Let $a_n = (1 + \frac{1}{n})^n (n \in \mathbb{Z}^+)$.
   Then $([1 + \frac{1}{2n}]^{2n})_{n=1}^{\infty}$ is the subsequence $(a_{2n})_{n=1}^{\infty}$ of $(a_n)$. Since the original sequence converges to e, every subsequence converges to the same limit e.

   Hence $$\lim_{n\to\infty} \left(1 + \frac{1}{2n}\right)^{2n} = e. \tag{1}$$

   Now $\left[\left(1 + \dfrac{1}{2n}\right)^n\right]^2 = \left(1 + \dfrac{1}{2n}\right)^{2n}$, and therefore, from (1),

   $$\lim_{n\to\infty} \left(1 + \frac{1}{2n}\right)^n = \sqrt{e}.$$
4. Let $(a_n)_{n=1}^{\infty}$ be a sequence of real numbers. Suppose we know that the subsequence $(a_{2n})_{n=1}^{\infty}$ converges to $a$, and the subsequence $(a_{2n+1})_{n=1}^{\infty}$ converges to $a$, then we can show that the original sequence $(a_n)$ also converges to $a$.

   Since $(a_{2n})$ converges to $a$, we know that given any $\varepsilon > 0$, there is some $N'$ such that

   $$|a_{2n} - a| < \varepsilon$$

   for all $n > N'$. However, $(a_{2n+1})$ also converges to $a$ and hence there is some

$N''$ such that

$$|a_{2n+1} - a| < \varepsilon$$

for all $n > N''$. Let $N = \max\{2N' + 1, 2N'' + 2\}$. Then for all $n > N$, we have

$$|a_n - a| < \varepsilon$$

and so $a_n \to a$ as $n \to \infty$.

## *EXERCISES 2.5.1*

**1** By using the fact that $(1 + 1/n)^n \to \mathrm{e}$ as $n \to \infty$, show that the following sequences are convergent and find their limits in terms of e:

$$\left(\left[1 + \frac{1}{3n}\right]^n\right), \quad \left(\left[1 - \frac{1}{n}\right]^n\right), \quad \left(\left[1 + \frac{1}{2n}\right]^n\right), \quad \left(\left[1 - \frac{1}{2n} - \frac{1}{2n^2}\right]^n\right).$$

**2** Let $(a_n)$ and $(b_n)$ be two convergent sequences of real numbers such that $a_n \to a$ as $n \to \infty$ and $b_n \to b$ as $n \to \infty$. Prove that, if $(a_n + (-1)^n b_n)$ and $((-1)^{n+1} a_n + b_n)$ both converge, then $a = b = 0$.

**3** Let $(a_n)_{n=1}^{\infty}$ be a sequence of real numbers. Suppose that each of the subsequences

$$(a_{3n})_{n=1}^{\infty}, \qquad (a_{3n+1})_{n=0}^{\infty}, \qquad (a_{3n+2})_{n=0}^{\infty}$$

converges to the limit $a$. Prove that the original sequence $(a_n)$ converges to $a$.

**4** Let $(a_n)_{n=1}^{\infty}$ be a sequence of real numbers. Let $p$ be any positive integer such that $p \geqslant 2$. Suppose each of the $p$ subsequences $(a_{pn})_{n=1}^{\infty}$, $(a_{pn+1})_{n=0}^{\infty}$, $(a_{pn+2})_{n=0}^{\infty}, \ldots, (a_{pn+p-1})_{n=0}^{\infty}$ converges to the limit $a$. Prove that the original sequence $(a_n)$ converges to $a$. (The reader should memorise this result as it is used in the next chapter—see Theorem 3.3.6 on the alternating series test and the Appendix on rearrangements.)

The definition of convergence for a sequence $(a_n)$ involved an investigation of $|a_n - a|$, where $a$ is the limit of the sequence. We therefore have a criterion for deciding whether $(a_n)$ converges to $a$, but we have no means, at present, of deciding whether a sequence converges if we do not know or are unable to guess a suitable value for the limit. What we need is a criterion which involves only the actual terms of the sequence and makes no mention of the value of the limit. The Cauchy criterion for convergence provides the answer. In the process of leading up to this proof we will need to show that every infinite sequence has a monotonic subsequence—a result which is of interest in its own right.

THEOREM 2.5.4 Every infinite sequence of real numbers has a monotonic subsequence.

*Proof* Let $(a_n)$ be an infinite sequence of real numbers. We call $a_p$ a terrace point if $a_p \geqslant a_n$ for all $n \geqslant p$. The sequence either has an infinite number of terrace points or it does not.

(a) If $(a_n)$ has an infinite number of terrace points, these can be labelled $a_{n_1}, a_{n_2}, a_{n_3}, \ldots$ where $n_1 < n_2 < n_3 < \ldots$. Since $a_{n_1}$ is a terrace point, $a_{n_1} \geqslant a_n$ for all $n \geqslant n_1$ and so in particular $a_{n_1} \geqslant a_{n_2}$. Similarly, $a_{n_2} \geqslant a_{n_3}$, $a_{n_3} \geqslant a_{n_4}$ and so on. Thus $(a_{n_k})_{k=1}^{\infty}$ is a decreasing subsequence of $(a_n)$.

(b) If $(a_n)$ does not have an infinite number of terrace points, then it only has a finite number (which may be zero). We can therefore find $n_1$ so that all the terrace points $a_p$ occur for values of $p < n_1$. Since $a_{n_1}$ is not a terrace point, there exists $n_2$ with $n_2 > n_1$ such that $a_{n_2} > a_{n_1}$. Now $a_{n_2}$ is not a terrace point and so there is an integer $n_3 > n_2$ such that $a_{n_3} > a_{n_2}$. Continuing in this way gives a sequence $a_{n_1}, a_{n_2}, a_{n_3}, \ldots$ such that $a_{n_1} < a_{n_2} < a_{n_3} < \ldots$. Thus $(a_{n_k})_{k=1}^{\infty}$ is a strictly increasing subsequence of $(a_n)$.

Cases (a) and (b) together show that $(a_n)$ has a monotonic subsequence.

COROLLARY Every bounded infinite sequence has a convergent subsequence.

*Proof* A bounded infinite sequence has a bounded monotonic subsequence. This subsequence is convergent by Theorem 2.4.3.

With the help of this result we can now deduce Cauchy's criterion for convergence.

THEOREM 2.5.5 (Cauchy's Criterion for Convergence)

Let $(a_n)$ be a sequence of real numbers with the property that for every $\varepsilon > 0$ there is corresponding $N$ such that

$$|a_n - a_m| < \varepsilon \qquad \text{for all } m, n > N. \tag{1}$$

Then $(a_n)$ is convergent.

*Proof* From (1) we see that there is an integer $M$ such that $|a_n - a_m| < 1$ for all $m, n > M$. In particular, $|a_n - a_{M+1}| < 1$ for all $n > M$. Hence

$$|a_n| = |a_n - a_{M+1} + a_{M+1}| \leqslant |a_n - a_{M+1}| + |a_{M+1}| < 1 + |a_{M+1}|$$

for all $n > M$. Let

$$K = \max\{|a_1|, |a_2|, |a_3|, \ldots, |a_{M-1}|, |a_M|, 1 + |a_{M+1}|\}.$$

Then $|a_n| \leqslant K$ for all $n$ and the sequence $(a_n)$ is bounded. By the corollary to the previous theorem $(a_n)$ has a convergent subsequence. Let this subsequence be $(a_{n_k})_{k=1}^{\infty}$ and suppose $a_{n_k} \to a$ as $k \to \infty$. Let $\varepsilon > 0$ be any positive real number; then there exist positive integers $N_0$, $K_0$ such that

$$|a_n - a_m| < \varepsilon/2 \qquad \text{for all } m, n > N_0 \tag{1}$$

and

$$|a_{n_k} - a| < \varepsilon/2 \qquad \text{for all } k > K_0. \tag{2}$$

Let $N = \max\{N_0, K_0\}$. Since $n_k \geqslant k$, we see that for all $k > N$

$$\begin{aligned} |a_k - a| &= |a_k - a_{n_k} + a_{n_k} - a| \leqslant |a_k - a_{n_k}| + |a_{n_k} - a| \\ &< \varepsilon/2 + \varepsilon/2 = \varepsilon \end{aligned}$$

using (1) and (2). Hence $a_k \to a$ as $k \to \infty$ and the sequence $(a_n)$ converges.

The converse is, of course, relatively easy to prove. Suppose the sequence $(a_n)$ converges to $a$. Then given $\varepsilon > 0$, there is a corresponding $N$ such that

$$|a_n - a| < \varepsilon/2 \qquad \text{for all } n > N.$$

Thus if $n, m > N$,

$$|a_n - a_m| = |a_n - a + a - a_m| \leqslant |a_n - a| + |a - a_m| < \varepsilon.$$

# APPENDIX: TRIANGLE INEQUALITIES

Expressions involving both moduli and inequalities are common in analysis. For example, we have the requirement $|a_n - a| < \varepsilon$ in the definition of a limit. The reader, therefore, cannot avoid inequalities and would be wise to learn how to handle them.

In analysis, the triangle inequalities prove invaluable. They are

$$\big||x| - |y|\big| \leqslant |x + y| \leqslant |x| + |y|, \tag{1}$$

$$\big||x| - |y|\big| \leqslant |x - y| \leqslant |x| + |y|, \tag{2}$$

for all real numbers $x, y$. In order to prove them we recall that $|x| = x$ if $x \geqslant 0$ and $|x| = -x$ if $x < 0$. Hence

$$|x|^2 = x^2 \tag{3}$$

for all real $x$, and therefore

$$|x| = +\sqrt{x^2}.$$

Now for all real numbers $x$,

$$x \leqslant +\sqrt{x^2}, \qquad -x \leqslant +\sqrt{x^2},$$

i.e. $$x \leqslant |x|; \qquad -x \leqslant |x|. \tag{4}$$

Thus if $x, y$ are any two real numbers, we have

$$|x+y|^2 = (x+y)^2 = x^2 + 2xy + y^2 = |x|^2 + 2xy + |y|^2$$

by (3) and therefore

$$|x+y|^2 \leqslant |x|^2 + 2|xy| + |y|^2 = (|x| + |y|)^2$$

from (4). Taking positive square roots gives

$$|x+y| \leqslant |x| + |y|. \tag{5}$$

Similarly, using (4) we have

$$|x+y|^2 = |x|^2 + 2xy + |y|^2 \geqslant |x|^2 - 2|xy| + |y|^2 = (|x| - |y|)^2.$$

Again take the positive square root to obtain

$$|x+y| \geqslant \big||x| - |y|\big|. \tag{6}$$

The outer modulus on the right-hand side of (6) occurs as there is no information about whether $|y|$ is larger than $|x|$ or not. Inequalities (5) and (6) together give the triangle inequality (1). Now that one of the triangle inequalities has been established there is no problem in proving the other. We simply need to note that $|-y| = |y|$, and the result follows immediately.

How do we use the triangle inequalities? Each time we have to decide which part of the inequalities to use. There is one golden rule governing the choice – **select the part of the inequality which gives the inequality sign the required way** for the current problem. Remember also that if $0 \leqslant a \leqslant b$ and $0 \leqslant c \leqslant d$, then $ac \leqslant bd$, but these inequalities give no information about the relative sizes of $a/c$ and $b/d$. Thus **inequalities may be multiplied, but they cannot be divided**. Let us try some examples.

---

### *Examples*

1. Let $a, b, c, d$ be real numbers such that $|c| \neq |d|$. Then

$$0 < \big||c| - |d|\big| \leqslant |c+d| \leqslant |c| + |d|,$$

and therefore

$$0 < \frac{1}{|c|+|d|} \leqslant \frac{1}{|c+d|} \leqslant \frac{1}{\big||c|-|d|\big|}. \tag{*}$$

Since $$0 \leqslant \big||a| - |b|\big| \leqslant |a+b| \leqslant |a| + |b|$$

and inequalities can be multiplied, we have

$$\frac{\big||a|-|b|\big|}{|c|+|d|} \leqslant \frac{|a+b|}{|c+d|} = \left|\frac{a+b}{c+d}\right| \leqslant \frac{|a|+|b|}{\big||c|-|d|\big|}.$$

The reader should pay particular attention to the denominators. The signs have been determined by inequality (*). It is the choice of the correct expression for the denominator which frequently causes confusion—so beware!

2. If $|x| \leqslant 1$, then $$|x-3| \leqslant 4$$
and $$|x^2-3| \geqslant \big||x^2|-|3|\big| = 3-|x|^2 \geqslant 3-1=2.$$
Hence $$\frac{1}{|x^2-3|} \leqslant \frac{1}{2}$$
and therefore $$\left|\frac{x-3}{x^2-3}\right| = \frac{|x-3|}{|x^2-3|} \leqslant \frac{4}{2} = 2$$
whenever $|x| \leqslant 1$.

## TABLE OF STANDARD LIMITS

**1 (2.3.1)** As $n \to \infty$, $n^\alpha \to \begin{cases} \infty & (\alpha > 0), \\ 1 & (\alpha = 0), \\ 0 & (\alpha < 0). \end{cases}$

**2 (2.3.2)** As $n \to \infty$, $a^n \to \begin{cases} \infty & (a > 1), \\ 1 & (a = 1), \\ 0 & (-1 < a < 1). \end{cases}$

If $a \leqslant -1$, then $a^n$ does not tend to any limit as $n \to \infty$.

**3 (2.3.3)** If $\alpha > 0$ and $a > 1$, $\dfrac{a^n}{n^\alpha} \to \infty$ as $n \to \infty$.

If $\alpha > 0$ and $|b| < 1$ then $n^\alpha b^n \to 0$ as $n \to \infty$.

**4 (2.3.4)** For $a > 0$, $a^{1/n} \to 1$ as $n \to \infty$.

**5 (2.3.5)** As $n \to \infty$, $n^{1/n} \to 1$.

**6 (2.3.6)** For all real numbers $a$, $\dfrac{a^n}{n!} \to 0$ as $n \to \infty$.

**7 (2.3.7)** As $n \to \infty$, $\dfrac{n!}{n^n} \to 0$.

**8** As $n \to \infty$, $\left(1 + \dfrac{x}{n}\right) \to e^x$ for every real number $x$. In particular,

$$\left(1+\frac{1}{n}\right)^n \to e$$

as $n \to \infty$.

## MISCELLANEOUS EXERCISES 2

**1** In each of the following cases, decide whether $a_n$ tends to a limit as $n \to \infty$. When the limit exists find it. Give reasons for your answers.

(a) $a_n = \dfrac{n^2 10^n + n^3 9^n}{7^{2n}+1}$.

(b) $a_n = \dfrac{n^2+7}{2+n^3}$.

(c) $a_n = \dfrac{2^n \cdot n^2 + 3^n}{3^n + n^{100}}$.

(d) $a_n = (3n^2+n)^{1/n}$.

(e) $a_n = \dfrac{2n^3+3n+1}{3n^3+4}$.

(f) $a_n = \dfrac{2n^3+3n+1}{3n^2+4}$.

(g) $a_n = \dfrac{6^n+n!}{n!+(7)^{2n}}$.

(h) $a_n = \dfrac{(n+1)^n}{n^n}$.

(i) $a_n = \dfrac{3^n+(-2)^n}{1+n!}$.

**2** For $x \geqslant 0$, let

$$a_n(x) = (1+x^n)^{1/n}.$$

Prove that if $0 \leqslant x \leqslant 1$, then $a_n(x) \to 1$ as $n \to \infty$.
Show also that if $x > 1$, then $a_n(x) \to x$ as $n \to \infty$.

**3** The sequence $(a_n)$ is defined by

$$a_1 = 1,\ a_{n+1} = \sqrt{1 + a_n^2/2} \qquad (n = 1, 2, 3, \ldots).$$

Show that (a) $a_n^2 - 2 < 0$, (b) $a_{n+1}^2 - a_n^2 > 0$.
Deduce that $(a_n)$ converges and find its limit.

**4** For each of the following sequences $(a_n)$, decide whether $a_n$ tends to a limit as $n \to \infty$. When the limit exists, find it.

(a) $a_n = [4^{10} + (\frac{1}{2})^n]^{1/n}$;

(b) $a_n = (4^{10} + 2^n)^{1/n}$;

(c) $a_n = \dfrac{3n^3+1}{n^3+n^2}$;

(d) $a_n = \dfrac{3n^3 + n\cos^2 n}{n^3 + \sin^2 n}$;

(e) $a_n = \dfrac{3n^3 + n\cos^2 n}{n^2 + \sin^2 n}$;

(f) $a_n = \dfrac{n^6 6^n + n^4 4^n}{n^5 5^n + 7^n}$;

(g) $a_n = n(\sqrt{n^2+144} - \sqrt{n^2-1})$.

Give reasons for your answers.

**5** Give an example of sequences $(a_n)$, $(b_n)$ which do not converge for which $[a_n + (-1)^n b_n]$ and $[(-1)^{n+1} a_n + b_n]$ both converge. (See Exercises 2.5.1, Question 2.)

## HINTS FOR SOLUTION OF EXERCISES

### Exercises 2.3.2

The algebra of limits is used in many of these parts. In addition, the following are also used:

**1** Use 2.3.3. **2** Use 2.3.3. **4** Use 2.3.3

**5** Use Theorem 2.2.1 and the identity $a^2 - b^2 = (a-b)(a+b)$. This allows $1/(\sqrt{n^2+2} - \sqrt{n^2+1})$ to be expressed as $\sqrt{n^2+2} + \sqrt{n^2+1}$.

**6** Use 2.3.2. **10** Use Theorem 2.2.1.

**11** and **12** Use 2.3.4 and the sandwich rule.

### Exercises 2.5.1

**1** $[(1+1/3n)^{3n}]$ is a subsequence of the original sequence and so has the same limit;

$$\left(1-\frac{1}{n}\right)^n = \left(\frac{n-1}{n}\right)^n = \frac{1}{\left(1+\dfrac{1}{n-1}\right)^n}$$

and
$$\left(1+\frac{1}{n-1}\right)^{n-1} \to \mathrm{e} \quad \text{as } n \to \infty;$$

$$\left(1-\frac{1}{2n}-\frac{1}{2n^2}\right) = \left(1-\frac{1}{n}\right)\left(1+\frac{1}{2n}\right).$$

### Miscellaneous Exercises 2

**4** (g) Use the identity $a^2 - b^2 = (a+b)(a-b)$ with $a = \sqrt{n^2+144}$, $b = \sqrt{n^2-1}$.

## ANSWERS TO EXERCISES

### Exercises 2.2.1

**1** $N$ is the smallest integer such that $N \geqslant \mathrm{e}^{\mathrm{e}^4}$.

## Exercises 2.3.2

**1** 1, **2** 0, **3** 1, **4** 0, **5** $\infty$, **6** 0, **7** $\infty$, **8** 0, **9** 3, **10** $\infty$, **11** 1, **12** 3, **13** $\frac{1}{4}$, **14** $(\frac{1}{3})^5$.

## Exercises 2.5.1

**1** $e^{1/3}$, $1/e$, $\sqrt{e}$, $1/\sqrt{e}$.

## Miscellaneous Exercises 2

**1** In each case $a_n$ tends to a limit as $n \to \infty$. The limits are given below, with reasons in brackets.

(a) 0 (algebra of limits and 2.3.3).
(b) 0 (algebra of limits). (c) 1 (2.3.3).
(d) 1 (sandwich rule and 2.3.5). (e) $\frac{2}{3}$ (algebra of limits).
(f) $\infty$ (theorem 2.2.1). (g) 1 (2.3.6).
(h) e (see section following Theorem 2.4.3).
(i) 0 (2.3.6).

**3** $\sqrt{2}$.

**4** In each case $a_n$ tends to a limit as $n \to \infty$. The limits are given below with reasons in brackets.

(a) 1 (sandwich rule and 2.3.4). (b) 2 (sandwich rule and 2.3.4).
(c) 3 (algebra of limits). (d) 3 (algebra of limits and sandwich rule).
(e) $\infty$ (theorem 2.2.1). (f) 0 (2.3.3). (g) $\frac{145}{2}$ (algebra of limits).

**5** One example is $a_n = 1 + (-1)^n$ and $b_n = -1 + (-1)^n$ for $n = 1, 2, 3, 4, \ldots$. In this example $a_n + (-1)^n b_n = 2$ and $(-1)^{n+1} a_n + b_n = -2$.

# 3 INFINITE SERIES

## 3.1 INTRODUCTION

In the previous chapter we investigated the behaviour of infinite sequences of real numbers and we realised that we had to look at the terms $a_n$ and consider the overall trend as $n$ gets larger and larger. For series, however, the important consideration is not just the individual terms themselves but rather the behaviour as we keep on adding them together.

As we remarked in the previous chapter, the idea of using infinite processes was first introduced by Newton in his treatise *De analysi per aequationes numero terminorum infinitas*. Newton is remembered as one of the outstanding mathematicians of all time. His work opened up numerous fields in mathematics, physics and astronomy and he developed new ideas which were to revolutionise mathematics. Among these ideas was the use of infinite series as a mathematical tool. It was left to succeeding generations of mathematicians to develop the subject rigorously and investigate the laws needed to govern the manipulation of infinite series. Newton and his contemporaries did not worry unduly about whether their use of the normal rules of algebra in dealing with infinite series would give the correct answer. They simply got on with their problems and on the whole their somewhat cavalier attitude seemed to work and produced many exciting results. Of course, they sometimes ran up against problems. For example, they were quite happy to use relations of the type

$$\frac{1}{1-x} = 1 + x + x^2 + x^3 + \dots. \tag{1}$$

Now if we substitute $x = 2$ in relation (1) then the l.h.s. has the value $-1$ and the r.h.s. reads $1 + 2 + 4 + 8 + 16 + \dots$. It would seem reasonable to write $\infty$ on the r.h.s. of (1). How can we square this with the fact that the l.h.s. has value $-1$? The answer is that we can't, but neither could the seventeenth-century mathematicians and so they just dismissed it as absurd. If, however, we substitute $x = \frac{1}{2}$ in relation (1), then the l.h.s. has value 2 and the r.h.s. is $1 + \frac{1}{2} + \frac{1}{4} + \frac{1}{8} + \dots$. Is the equality in relation (1) true for $x = \frac{1}{2}$? It certainly doesn't look absurd, but in view of what happened when we substituted $x = 2$, we remain a little cautious. We are in good company: the Swiss mathematician Leonhard Euler (1707–83) would also have advised caution. He recognised that it is unwise to use infinite series unless they are convergent, and he set the stage for a rigorous development of the subject. Rather surprisingly, he did not always exhibit caution in his own work involving infinite series. He did, however, produce many staggering results and used infinite series to effect in solving difficult and long-standing problems. His recognition of the underlying problem spurred others to develop the necessary theory which is part of present day analysis.

A prime mover in the development of this new theory was the German mathematician Carl Friedrich Gauss (1777–1855). Stories abound about his childhood and he was certainly an amazingly precocious infant. Perhaps the best-known story concerns his first arithmetic lesson at the age of nine. By all accounts his teacher, Büttner, was an unpleasant and sullen character. As the story goes, Büttner set his class a problem consisting of adding 100 numbers together. Whether these numbers were the first 100 positive integers, as some commentators think or the numbers $81\,297 + 81\,495 + 81\,693 + 81\,891 + \ldots + 100\,899$ as others claim is immaterial. The important point was that the 100 numbers formed an arithmetic progression which the teacher could readily sum by using a formula whereas the pupils in their first arithmetic lesson would normally only obtain the required answer after long and laborious additions. The school rule was that each pupil put his slate on the teacher's desk when he had completed the task. Nine-year old Gauss put his slate in front of the teacher almost immediately the teacher stopped speaking. The other pupils toiled on for about an hour. When the slates were finally checked only one pupil had the correct answer—the nine-year-old Gauss. Without prior knowledge of the formula for arithmetic progressions he had managed to work out how to do the computation, while the teacher spoke—a phenomenal achievement for a nine-year-old. Throughout his long life he remained capable of doing complicated arithmetic operations in his head. Fortunately, the fact that Gauss had outstanding ability was acknowledged and he was given the opportunity to develop it.

One of his early interests was the binomial expansion of $(1 + x)^n$ in the case in which $n$ is not a positive integer. Some so-called proofs had been devised, but young Gauss realised that they were unsatisfactory and set about making good the omission. Relation (1) which occurred earlier in this chapter is just a particular example of the binomial expansion for the case $n = -1$. We have seen that using the value $x = 2$ leads to an absurdity of the form: 'minus one equals infinity'. Gauss was the first to see that a 'proof' which can lead to such a ridiculous statement is no proof at all. He realised that the statement is correct for certain values, but not for all values of $x$ and $n$. In his opinion a formula which gave consistent results for some values had no place in mathematics until the precise conditions had been determined for which consistency is guaranteed. This criterion prescribes the form in which analytical theorems are presented right up to the present day. Typical theorems in analysis are of the form: 'if certain specified conditions are satisfied, then a given conclusion follows'.

We have Gauss to thank for this, but the man who more than any other put this part of analysis in a form which we recognise readily today is the French mathematician Cauchy (1789–1857). Not suprisingly, his name is now associated with many facets of analysis. For example, we have Cauchy's criterion for convergence. He was a very gifted teacher and a prolific writer. In 1821 he wrote for publication the course of lectures on analysis which he had been giving at the Polytéchnique. Much of what he wrote on convergence of infinite series in this course of lectures would not surprise a modern reader, as it is so like the accounts in present-day textbooks. In these lectures he introduced the idea of taking the sum $s_n$ of the first $n$ terms of the series and considering what happens to $s_n$ as $n \to \infty$, which is exactly the line we will follow in the next section.

Cauchy was a prolific writer—producing in his lifetime over 800 papers, mémoires, treatises, etc. (one of them being over 300 pages long), and his terrific activity had a rather amusing result. In 1835 the Academy of Sciences began publishing its weekly bulletin (the '*Comptes Rendus*') and Cauchy deluged the new publication with articles. In view of the rapidly rising printing bill, the Academy passed a rule (which is still in force today) prohibiting the publication of any paper over four pages long. Cauchy had to look for other places to publish his longer works.

As a lecturer Cauchy was apparently very effective. According to old stories one well-known French mathematician (cum astronomer), Laplace (1749–1827), is said to have listened to his lectures on infinite series with attention and mounting apprehension. Unlike many mathematicians over the ages, Laplace directed most of his energies towards one major project—the investigation of the stability of the solar system. His *magnum opus* was the *Mècanique cèleste*—published in five volumes over a period of 26 years from 1799 to 1825. In the course of his investigations Laplace had used infinite series. After listening to Cauchy's lecture, he rushed home to check whether the series he had used were convergent, fearing that all his work on celestial mechanics might be destroyed if his series turned out to be divergent. To his great relief, he found that his series were indeed convergent and his work was saved!

We too can use infinite series and use them with confidence provided we keep the laws governing their use. The purpose of this chapter is to develop these rules.

## 3.2 SERIES AND NOTATION

We first introduce the sigma notation. Let $a_1, a_2, a_3, \ldots$ be real numbers. When we need notation for the sum of the first $n$ terms $a_1 + a_2 + \ldots + a_n$, we will write $$\sum_{k=1}^{n} a_k = a_1 + a_2 + \ldots + a_n.$$

If we wish to keep on adding the terms indefinitely, giving $a_1 + a_2 + a_3 + \ldots$ (where the dots signify that we keep on adding more and more terms), then we will use

$$\sum_{n=1}^{\infty} a_n = a_1 + a_2 + a_3 + \ldots.$$

At the moment this is just a formal sum as we do not know whether it makes any sense to keep on adding more and more terms. In order to investigate further, it would seem sensible to find the sum of the first $n$ terms and then consider what happens to this sum as $n$ increases. Using the standard notation we would normally write

$$s_n = a_1 + a_2 + \ldots + a_n.$$

If $s_n$ tends to a finite limit $s$ as $n \to \infty$, then the sequence $(s_n)$ converges and it is sensible to say that the series $\sum_{n=1}^{\infty} a_n$ converges and call the limit $s$ its sum.

To facilitate printing, the limits are placed alongside the summation signs when they appear in a line of text. A neater appearance can be achieved by using $\sum_1^{\infty} a_n$ instead of $\sum_{n=1}^{\infty} a_n$ when there is no danger of confusion.

DEFINITION 3.2.1 Let $a_1, a_2, \ldots$ be real numbers and let

$$s_n = a_1 + a_2 + \ldots a_n = \sum_{k=1}^{n} a_k.$$

The infinite series $\sum_1^{\infty} a_n$ is said to **converge** if $s_n$ tends to a finite limit as $n \to \infty$. The value of this finite limit is called the **sum of the series** and we write

$$\sum_{n=1}^{\infty} a_n = s,$$

where $s = \lim_{n \to \infty} s_n$. The quantity $s_n$ is called a **partial sum** of the infinite series.

DEFINITION 3.2.2 An infinite series

$$\sum_{n=1}^{\infty} a_n$$

which does not converge is said to **diverge**.

The symbols $$\sum_{n=1}^{\infty} a_n$$

have been used in two different senses. In the first case they represent the formal sum $a_1 + a_2 + a_3 + a_4 + \ldots$. In the definition, however, they are also used for the value of the sum of a convergent series and the equality

$$\sum_{n=1}^{\infty} a_n = s$$

has a well-defined meaning. Normally, it will be clear from the context how

$$\sum_{n=1}^{\infty} a_n$$

is being used and so the two different meanings do not cause any confusion.

---

*Examples 3.2.1*

1. Consider the geometric series $1 + x + x^2 + \ldots$ Write

$$s_n = 1 + x + x^2 + \ldots + x^{n-1},$$

i.e. $s_n$ is the sum of the first $n$ terms. Then

$$xs_n = x + x^2 + \ldots + x^{n-1} + x^n,$$

and, therefore $$(1-x)s_n = 1 - x^n.$$

Thus if $x \neq 1$, $s_n = \dfrac{1-x^n}{1-x}$.
If $x = 1$, $s_n = 1 + 1 + \ldots + 1 = n$.
Now if $|x| < 1$, $x^n \to 0$ as $n \to \infty$ and hence

$$s_n \to \frac{1}{1-x}$$

as $n \to \infty$. The series

$$\sum_{n=0}^{\infty} x^n,$$

therefore, converges for $|x| < 1$ and its sum is $1/(1-x)$.
If $x = 1$, $s_n = n$ and $s_n \to \infty$ as $n \to \infty$.
If $x = -1$, $s_n = \dfrac{1-(-1)^n}{2}$ and $s_n$ does not tend to any limit as $n \to \infty$.
If $|x| > 1$, then

$$|s_n| = \frac{|x^n - 1|}{|x - 1|} \geqslant \frac{|x|^n - 1}{|x| + 1}$$

and $|s_n| \to \infty$ as $n \to \infty$. Hence $s_n$ cannot tend to a finite limit as $n \to \infty$ if $|x| > 1$. Hence the geometric series

$$\sum_{n=0}^{\infty} x^n$$

diverges if $|x| \geqslant 1$ and it converges if $|x| < 1$. Moreover, for $|x| < 1$

$$\sum_{n=0}^{\infty} x^n = \frac{1}{1-x}.$$

2. Let $a_n = \dfrac{1}{n(n+1)}$ $(n \in \mathbb{Z}^+)$. Then

$$a_n = \frac{1}{n} - \frac{1}{n+1}.$$

Hence

$$\begin{aligned} a_1 &= 1 - \tfrac{1}{2} \\ a_2 &= \tfrac{1}{2} - \tfrac{1}{3} \\ a_3 &= \tfrac{1}{3} - \tfrac{1}{4} \\ a_{n-1} &= \frac{1}{n-1} - \frac{1}{n} \\ a_n &= \frac{1}{n} - \frac{1}{n+1} \end{aligned}$$

and it follows that

$$s_n = a_1 + a_2 + \ldots a_n = 1 - \frac{1}{n+1}.$$

Thus $s_n \to 1$ as $n \to \infty$, the series

$$\sum_{n=1}^{\infty} a_n$$

converges and

$$\sum_{n=1}^{\infty} \frac{1}{n(n+1)} = 1.$$

3. Let

$$\sum_{n=1}^{\infty} a_n \quad \text{and} \quad \sum_{n=1}^{\infty} b_n$$

be convergent series. Write

$$\begin{aligned} s_n &= a_1 + a_2 + \ldots + a_n, \\ t_n &= b_1 + b_2 + \ldots + b_n. \end{aligned}$$

Then $$(a_1+b_1)+(a_2+b_2)+\ldots+(a_n+b_n)=s_n+t_n$$

and $s_n+t_n \to s+t$ as $n\to\infty$, where

$$s=\sum_{n=1}^{\infty} a_n \quad \text{and} \quad t=\sum_{n=1}^{\infty} b_n.$$

Hence $\sum_{n=1}^{\infty}(a_n+b_n)$ converges and

$$\sum_{n=1}^{\infty}(a_n+b_n)=\sum_{n=1}^{\infty} a_n+\sum_{n=1}^{\infty} b_n.$$

There is therefore no difficulty about adding convergent series. We shall see later than multiplication of series is not quite so straightforward.

## *EXERCISES 3.2.1*

**1** Let $a_n=1/[n(n+1)(n+2)]$ $(n\in\mathbb{Z}^+)$ and let $s_n=a_1+a_2+\ldots+a_n$. Express $a_n$ in partial fractions and hence, or otherwise, show that

$$s_n=\frac{1}{4}-\frac{1}{2(n+1)}+\frac{1}{2(n+2)}.$$

Deduce that

$$\sum_{n=1}^{\infty} 1/[n(n+1)(n+2)]$$

converges and find its sum.

**2** Find an explicit expression for the partial sums $s_n$ of

$$\sum_{n=1}^{\infty} \log\left(1+\frac{1}{n}\right).$$

Deduce that

$$\sum_{n=1}^{\infty} \log\left(1+\frac{1}{n}\right)$$

diverges.

In two of the worked examples, it was possible to write down an explicit expression for the partial sum $s_n$ and to then check whether $s_n$ tends to a finite limit as $n\to\infty$. It rarely happens that this can be done in the general case and so we must build up a collection of tests which can be used to decide whether a series converges. The first test is a very simple one.

## 3.3 TESTS FOR CONVERGENCE

In a sense, the first test is probably better considered as a test for divergence.

THEOREM 3.3.1 If $a_n \nrightarrow 0$ as $n \to \infty$, then

$$\sum_{n=1}^{\infty} a_n$$

diverges.

*Proof* Write $$s_n = a_1 + a_2 + \ldots + a_n.$$

If the given series converges then $s_n$ tends to a finite limit $s$ as $n \to \infty$. Hence $s_{n-1} \to s$ as $n \to \infty$, and therefore

$$a_n = s_n - s_{n-1} \to s - s = 0$$

as $n \to \infty$. It follows that if $a_n \nrightarrow 0$, then $s_n$ cannot tend to a finite limit as $n \to \infty$ and the series must diverge.

COROLLARY If $\sum a_n$ converges, then $a_n \to 0$ as $n \to \infty$.

The reader is warned to be extremely careful about the use of this corollary. It is impossible to have convergence unless $a_n \to 0$ as $n \to \infty$, but $a_n \to 0$ as $n \to \infty$ is not sufficient to guarantee convergence. For there are divergent series $\sum a_n$ for which $a_n \to 0$ as $n \to \infty$.

---

*Example 3.3.1*

Let $a_n = \frac{1}{n}$ $(n \in \mathbb{Z}^+)$. Then $a_n \to 0$ as $n \to \infty$. Let

$$s_n = a_1 + a_2 + \ldots + a_n = 1 + \tfrac{1}{2} + \tfrac{1}{3} + \tfrac{1}{4} + \ldots + \tfrac{1}{n}.$$

We will show that $s_n \to \infty$ as $n \to \infty$ and so $\sum a_n$ diverges despite the fact that $a_n \to 0$ as $n \to \infty$. To achieve this first let us bracket the terms in groups in the following way:

$$1 + \frac{1}{2} + \left(\frac{1}{3} + \frac{1}{4}\right) + \left(\frac{1}{5} + \frac{1}{6} + \frac{1}{7} + \frac{1}{8}\right) + \ldots + \left(\frac{1}{2^{n-1}+1} + \ldots + \frac{1}{2^n}\right) + \ldots,$$

so that the final term in each bracket is the reciprocal of a power of two. Consider the sum of the first $2^n$ terms,

$$s_{2^n} = 1 + \frac{1}{2} + \left(\frac{1}{3} + \frac{1}{4}\right) + \left(\frac{1}{5} + \frac{1}{6} + \frac{1}{7} + \frac{1}{8}\right) + \ldots + \left(\frac{1}{2^{n-1}+1} + \ldots + \frac{1}{2^n}\right)$$

$$\geqslant 1+\frac{1}{2}+\left(\frac{1}{4}+\frac{1}{4}\right)+\left(\frac{1}{8}+\frac{1}{8}+\frac{1}{8}+\frac{1}{8}\right)+\ldots+\left(\frac{1}{2^n}+\frac{1}{2^n}+\ldots+\frac{1}{2^n}\right)$$
$$=1+\frac{1}{2}+\frac{1}{2}+\frac{1}{2}+\ldots+\frac{1}{2}$$
$$=1+(n)\frac{1}{2}=\frac{n+2}{2},$$

since each of the brackets in the second line has sum $\frac{1}{2}$. As each of the terms is positive we see that $(s_n)$ is an increasing sequence. Moreover,

$$s_{2^n}\geqslant\frac{n+1}{2},$$

and so $(s_n)$ is not bounded above. Hence $s_n\to\infty$ as $n\to\infty$ by Theorem 2.4.1. Thus the series

$$\sum_{n=1}^{\infty}\frac{1}{n}$$

diverges.

This result may seem a little surprising at first sight. It was not immediately obvious that we could make $s_n$ arbitrarily large by taking a sufficiently large number of terms, but it is indeed true. In fact the method we have used demonstrates that we can be certain $s_n > 10^6$ for all $n > 2^{1\,999\,998}$, which is, of course, rather a large number of terms.

This example should be remembered; it is a divergent series which is frequently used for reference.

---

Exercises 3.3.1 provide other examples of divergent infinite series $\sum a_n$ for which $a_n\to 0$ as $n\to\infty$.

---

*Examples 3.3.2*

1. Let
$$a_n=\left(\frac{3n+1}{5n+1}\right)^4.$$

Then
$$a_n=\left(\frac{3+\dfrac{1}{n}}{5+\dfrac{1}{n}}\right)^4\to\left(\frac{3}{5}\right)^4 \quad \text{as } n\to\infty$$

and
$$\sum_{n=1}^{\infty}a_n$$

diverges because $a_n\nrightarrow 0$ as $n\to\infty$.

2. Let $$a_n = \frac{1}{\sqrt{n}}.$$

Then $$a_n = \frac{1}{\sqrt{n}} \geqslant \frac{1}{n} \quad \text{for all } n. \tag{1}$$

Write $$s_n = a_1 + a_2 + a_3 + \ldots + a_n$$

$$t_n = b_1 + b_2 + \ldots + b_n,$$

where $b_n = \frac{1}{n}$. Then $$s_n \geqslant t_n$$

for all $n$ by (1) and so $s_n \to \infty$ as $n \to \infty$ because $t_n \to \infty$ as $n \to \infty$ i.e.

$$\sum_{n=1}^{\infty} a_n$$

diverges.

---

The reader will notice that the series

$$\sum_{n=1}^{\infty} (1/\sqrt{n})$$

was shown to diverge by using an inequality which compared its terms with the terms of the series

$$\sum_{n=1}^{\infty} 1/n$$

which is known to diverge. This suggests that we may be able to test for convergence or divergence by comparing the individual terms with those of a series whose behaviour is already known. Indeed, such a method is embodied in the well-known comparison test.

THEOREM 3.3.2 (Comparison Test)

Let $$\sum_{n=1}^{\infty} a_n \quad \text{and} \quad \sum_{n=1}^{\infty} b_n$$

be two series of non-negative real numbers, such that

(a) $a_n \leqslant K b_n$ for all $n$ and some positive real number $K$, and

(b) $\sum_{n=1}^{\infty} b_n$ converges.

Then $\sum_{n=1}^{\infty} a_n$ also converges.

*Proof* Let

$$s_n = a_1 + a_2 + \ldots + a_n,$$
$$t_n = b_1 + b_2 + \ldots + b_n.$$

Since $a_n \geqslant 0$ and $b_n \geqslant 0$ for all $n$, the sequences $(s_n)$ and $(t_n)$ are both increasing. Moreover, $t_n$ tends to a finite limit $t$ as $n \to \infty$, because

$$\sum_{n=1}^{\infty} b_n$$

is convergent. Hence $t_n \leqslant t$ for all $n$. Now

$$s_n = a_1 + a_2 + \ldots + a_n \leqslant K(b_1 + b_2 + \ldots + b_n) = Kt_n \leqslant Kt$$

for all $n$ and, therefore, $(s_n)$ is bounded above. Since $(s_n)$ is also increasing, we see that $s_n$ tends to a finite limit as $n \to \infty$, and

$$\sum_{n=1}^{\infty} a_n$$

therefore converges.

COROLLARY 1

Let

$$\sum_{n=1}^{\infty} a_n, \qquad \sum_{n=1}^{\infty} c_n$$

be two series of non-negative real numbers such that

(a) $a_n \geqslant kc_n$ for all $n$ and some positive real number $k$,
(b) $\sum_{n=1}^{\infty} c_n$ diverges.

Then

$$\sum_{n=1}^{\infty} a_n$$

also diverges.

*Proof* Write

$$s_n = a_1 + a_2 + \ldots + a_n,$$
$$r_n = c_1 + c_2 + \ldots + c_n.$$

From condition (a) it follows that

$$s_n \geqslant kr_n \tag{1}$$

for all $n$. Now $(r_n)$ is an increasing sequence which does not converge by (b). Hence $r_n \to \infty$ as $n \to \infty$ and, therefore, $kr_n \to \infty$ as $n \to \infty$. It now follows from (1) that $s_n \to \infty$ as $n \to \infty$ and so $\sum_{n=1}^{\infty} a_n$ diverges. This completes the proof.

It is not essential that the inequality (1) should be satisfied for the first few terms. In fact it is sufficient if this inequality is satisfied for all $n \geqslant N_0$ where $N_0$ is a given positive integer. Only a slight modification of the proof is needed to take into account this change. Thus we have the following corollaries.

COROLLARY 2

Let $$\sum_{n=1}^{\infty} a_n \quad \text{and} \quad \sum_{n=1}^{\infty} b_n$$

be two series of non-negative real numbers such that

(a) $a_n \leqslant Kb_n$ for all $n \geqslant N_0$ and some positive real number $K$, where $N_0$ is a given positive integer; and

(b) $\sum_{n=1}^{\infty} b_n$ converges.

Then $$\sum_{n=1}^{\infty} a_n$$

also converges.

COROLLARY 3

Let $$\sum_{n=1}^{\infty} a_n \quad \text{and} \quad \sum_{n=1}^{\infty} c_n$$

be two series of non-negative real numbers such that

(a) $a_n \geqslant Kc_n$ for all $n \geqslant N_0$ and some positive real number $K$, where $N_0$ is a given positive integer; and

(b) $\sum_{n=1}^{\infty} c_n$ diverges.

Then $$\sum_{n=1}^{\infty} a_n$$

also diverges.

Inevitably, mistakes will be made unless inequalities are handled with care. In some instances, however, the use of inequalities can be avoided by the use of an alternative form of the comparison test. This alternative is called the limit form of the comparison test.

THEOREM 3.3.2(a) (Limit Form of Comparison Test)

Let $$\sum_{n=1}^{\infty} a_n \quad \text{and} \quad \sum_{n=1}^{\infty} b_n$$

be series of **positive** real numbers such that $(a_n/b_n)$ tends to a **finite non-zero limit** as $n \to \infty$. Then either

$$\sum_{n=1}^{\infty} a_n \quad \text{and} \quad \sum_{n=1}^{\infty} b_n \tag{1}$$

both converge or

$$\sum_{n=1}^{\infty} a_n \quad \text{and} \quad \sum_{n=1}^{\infty} b_n \tag{2}$$

both diverge.

*Proof* Let the finite non-zero limit be $l$. Since $l \neq 0$ and $(a_n/b_n) \to l$ as $n \to \infty$, it follows that $l > 0$ and there is some integer $N$ such that

$$\left|\frac{a_n}{b_n} - l\right| < \tfrac{1}{2}l \qquad \text{for all } n > N,$$

i.e.

$$\tfrac{1}{2}l < \frac{a_n}{b_n} < \tfrac{3}{2}l \qquad \text{for all } n > N,$$

which gives

$$\tfrac{1}{2}lb_n < a_n < \tfrac{3}{2}lb_n \qquad \text{for all } n > N. \tag{1}$$

Now if $\sum_{n=1}^{\infty} b_n$ converges, then $\sum_{n=1}^{\infty} a_n$ must also converge by the comparison test since $0 < a_n < \frac{3}{2}lb_n$ for all $n > N$. Moreover, if $\sum_{n=1}^{\infty} a_n$ converges then $\sum_{n=1}^{\infty} b_n$ must also converge by the comparison test since $0 < b_n < 2a_n/l$ for all $n > N$. Thus if one of the series converges, then the other also converges, i.e. either

$$\sum_{n=1}^{\infty} a_n \quad \text{and} \quad \sum_{n=1}^{\infty} b_n \tag{1}$$

both converge or

$$\sum_{n=1}^{\infty} a_n \quad \text{and} \quad \sum_{n=1}^{\infty} b_n \tag{2}$$

both diverge.

---

*Examples 3.3.3*

1. Let $a_n = 1/n^2$, and $b_n = 1/[n(n+1)]$. Then for all $n \geqslant 1$, $a_n > 0$, $b_n > 0$

   and
   $$\frac{a_n}{b_n} = \frac{n(n+1)}{n^2} = \left(1 + \frac{1}{n}\right) \to 1 \qquad \text{as } n \to \infty.$$

   Now the second example of section 3.2 shows that $\sum_{n=1}^{\infty} 1/[n(n+1)]$ converges, and therefore, by the limit from the comparison test, $\sum_{n=1}^{\infty} 1/n^2$ converges.

2. Let $a_n = 1/n^\alpha$ where $\alpha \geqslant 2$ and $b_n = 1/n^2$. For all $n \in \mathbb{Z}^+$, $a_n > 0$, $b_n > 0$ and $0 < a_n = 1/n^\alpha \leqslant 1/n^2 = b_n$. Since $\sum_{n=1}^\infty b_n$ converges it follows by the comparison test that $\sum_{n=1}^\infty a_n$ converges.

Hence
$$\boxed{\sum_{n=1}^{\infty} \frac{1}{n^\alpha} \quad \text{converges for all } \alpha \geqslant 2.}$$

3. Let $a_n = 1/n$, $b_n = 1/n^\alpha$ where $\alpha \leqslant 1$. For all $n \in \mathbb{Z}^+$, $a_n > 0$, $b_n > 0$ and

$$0 < a_n = \frac{1}{n} \leqslant \frac{1}{n^\alpha} = b_n$$

since $\alpha \leqslant 1$. By Corollary 1 of the comparison test, $\sum_{n=1}^\infty b_n$ diverges.

Thus
$$\boxed{\sum_{n=1}^{\infty} \frac{1}{n^\alpha} \quad \text{diverges for } \alpha \leqslant 1.}$$

4. Let $a_n = \sin(1/n)$, $b_n = 1/n$. For all $n \in \mathbb{Z}^+$,

$$0 < 1/n \leqslant 1 < \pi/2$$

and, therefore, $\sin(1/n) > 0$ for $n \in \mathbb{Z}^+$.

Thus $a_n > 0$, $b_n > 0$ and

$$\frac{a_n}{b_n} = \frac{\sin(1/n)}{1/n} \to 1 \qquad \text{as } n \to \infty.$$

Now $\sum_{n=1}^\infty b_n$ diverges, and therefore $\sum_{n=1}^\infty a_n$ also diverges by the limit form of the comparison test.

---

Examples 2 and 3 above provide useful standard results which can be used in conjunction with the comparison test. At the moment they leave a small gap. With the tests at present at our disposal we cannot determine what happens to $\sum_{n=1}^\infty 1/n^\alpha$ for $1 < \alpha < 2$. Later we will be able to show, with the help of the integral test, that the series converges for these values of $\alpha$ also.

## *EXERCISES 3.3.1*

**1** Show that the following series converge:

(a) $\displaystyle\sum_{n=1}^{\infty} \frac{n^3 - 1}{4n^5 - 3n^2 + 3}$; (b) $\displaystyle\sum_{n=1}^{\infty} \left(\frac{n+1}{n^2+1}\right)^2$;

(c) $\displaystyle\sum_{n=1}^{\infty} \frac{3^n + 4^n}{3^n + 5^n}$; (d) $\displaystyle\sum_{n=1}^{\infty} \frac{\sqrt{n^2+1} - \sqrt{n^2-1}}{n}$.

**2** Show that the following series diverge:

(a) $\sum_{n=1}^{\infty} \frac{n^3+2n}{4n^4+3n^3+3}$; (b) $\sum_{n=1}^{\infty} (\sqrt{n^2+1}-\sqrt{n^2-1})$;

(c) $\sum_{n=1}^{\infty} \frac{3^n+5^n}{4^n}$; (d) $\sum_{n=1}^{\infty} \sin\left(\frac{1}{n}\right)$; (e) $\sum_{n=1}^{\infty} \cos\left(\frac{1}{n}\right)$.

**3** Determine whether the following series converge ot diverge:

(a) $\sum_{n=1}^{\infty} \frac{n+1}{n^2+1}$; (b) $\sum_{n=1}^{\infty} \frac{1}{n}\sin\left(\frac{1}{n}\right)$;

(c) $\sum_{n=1}^{\infty} \sin^2\left(\frac{1}{n}\right)$; (d) $\sum_{n=1}^{\infty} \frac{\sqrt{n+1}-\sqrt{n-1}}{\sqrt{n}}$;

(e) $\sum_{n=1}^{\infty} \frac{1}{n^{(n+1/n)}}$; (f) $\sum_{n=1}^{\infty} \frac{3^{1/n}}{n^{1/3}}$.

Reasons should be given for answers. (In solving these problems, the reader must be prepared to use all the tests given so far in section 3.3. Not all the questions rely on the comparison test!)

Using the comparison test, we examine a series for convergence by comparing it with another series whose behaviour is already known. It is, however, possible to test for convergence by studying only the behaviour of the ratio of successive terms. This test is known either as the ratio test or as d'Alembert's test in honour of the celebrated French mathematician Jean le Rond d'Alembert (1717–83). D'Alembert, an illegitimate son of the Chevalier Destouches, took his name from the Chapel of St Jean le Rond on whose steps he was abandoned, as an infant, by his mother. His name is known to many generations of students in connection with the following result.

THEOREM 3.3.3 (D'Alembert's Test—Ratio Test)

Let $\sum_{n=1}^{\infty} a_n$ be a series of positive real numbers such that $a_{n+1}/a_n \to l$ as $n \to \infty$.

(a) If $l<1$, then $\sum_{n=1}^{\infty} a_n$ converges.
(b) If $l>1$, then $\sum_{n=1}^{\infty} a_n$ diverges.

*Proof*

(a) *Case $l<1$* Choose a number $k$ such that $l<k<1$. Since $a_{n+1}/a_n \to l$ as $n \to \infty$, there is some $N$ such that

$$\left|\frac{a_{n+1}}{a_n}-l\right|<k-l$$

for all $n > N$. This gives

$$\frac{a_{n+1}}{a_n} < k$$

for all $n > N$.

i.e. $$a_{n+1} < ka_n \qquad \text{for all } n > N.$$

Thus for all $n > N+1$

$$a_n < ka_{n-1} < k^2 a_{n-2} < \ldots < k^{n-N-1} a_{N+1} = k^n \left( \frac{a_{N+1}}{k^{N+1}} \right).$$

Now $a_{N+1}/k^{N+1}$ is just a constant. Moreover, $\sum_{n=1}^{\infty} k^n$ converges since $0 < k < 1$. By the comparison test, $\sum_{n=1}^{\infty} a_n$ converges also.

(b) *Case* $l > 1$ Since $a_{n+1}/a_n \to l > 1$ as $n \to \infty$, there is some $M$ such that

$$\frac{a_{n+1}}{a_n} > 1$$

for all $n > M$. Thus for $n > M+1$

$$a_n > a_{n-1} > a_{n-2} > \ldots > a_{M+1}.$$

Hence $a_n \not\to 0$ as $n \to \infty$, and therefore $\sum_{n=1}^{\infty} a_n$ diverges.

Quite clearly the argument in the preceding paragraph can be used if $a_{n+1}/a_n \to \infty$ as $n \to \infty$, which gives us the following corollary.

COROLLARY

Let $$\sum_{n=1}^{\infty} a_n$$

be a series of positive terms such that $a_{n+1}/a_n \to \infty$ as $n \to \infty$.

Then $$\sum_{n=1}^{\infty} a_n$$

diverges.

In the statement of the theorem one case is obviously omitted, viz. the case $l = 1$. There is good reason for this omission. For if $l = 1$, either convergence or divergence is possible. For instance, if $a_n = 1/n$, then $\sum_{n=1}^{\infty} a_n$ diverges, but $a_{n+1}/a_n = n/n+1 \to 1$ as $n \to \infty$. However, if $a_n = 1/n^2$ then $\sum_{n=1}^{\infty} a_n$ converges, but again $a_{n+1}/a_n \to 1$ as $n \to \infty$. In both these examples $0 < a_{n+1} < a_n$ and yet one of the series converges and the other diverges. It is therefore inadmissible to omit the limit from the ratio test—a mistake frequently made by undergraduates. The condition $a_{n+1}/a_n < 1$ is not sufficient to guarantee convergence, as we can readily see by looking at the divergent series $\sum_{n=1}^{\infty} 1/n$.

*Examples 3.3.4*

1. Let $a_n = [(2n)!]^2/(4n)!$ for all $n \in \mathbb{Z}^+$. Then $a_n > 0$ and

$$\frac{a_{n+1}}{a_n} = \frac{[(2n+2)!]^2(4n)!}{(4n+4)![(2n)!]^2}$$
$$= \frac{(2n+2)(2n+1)(2n+2)(2n+1)}{(4n+4)(4n+3)(4n+2)(4n+1)}$$
$$= \frac{(2n+2)(2n+1)}{2(4n+3)2(4n+1)} \to \frac{1}{16}$$

as $n \to \infty$. Since the limit is less than 1, $\sum_{n=1}^{\infty} a_n$ converges.

2. Let $a_n = n^2 e^{-n(n+1)}$ for all $n \in \mathbb{Z}^+$. Then $a_n > 0$ and

$$\frac{a_{n+1}}{a_n} = \frac{(n+1)^2}{n^2} \frac{e^{n(n+1)}}{e^{(n+1)(n+2)}} = \left(1 + \frac{1}{n}\right)^2 e^{-2(n+1)} \to 0$$

as $n \to \infty$. Hence $\sum_{n=1}^{\infty} a_n$ converges.

## *EXERCISES 3.3.2*

Give reasons to justify your answers to the following questions. (In question 3, the reader must be prepared to use all the tests in the preceding part of section 3.3.)

**1** Prove that the following series converge:

(a) $\sum_{n=1}^{\infty} \frac{(n!)^2}{(2n)!}$; (b) $\sum_{n=1}^{\infty} \frac{(2n)!}{7^n(n!)^2}$; (c) $\sum_{n=1}^{\infty} \frac{3^n}{4^n\sqrt{n}}$;

(d) $\sum_{n=1}^{\infty} \frac{\cosh n}{3^n}$.

**2** Prove that the following series diverge:

(a) $\sum_{n=1}^{\infty} \frac{5^n(n!)^2}{(2n)!}$; (b) $\sum_{n=1}^{\infty} \frac{(2n)!}{(n!)^2}$; (c) $\sum_{n=1}^{\infty} \frac{\cosh n}{n}$;

(d) $\sum_{n=1}^{\infty} \frac{3^n + 5^n}{n4^n}$.

**3** Determine whether the following series converge:

(a) $\sum_{n=1}^{\infty} \frac{3^n + 5^n}{4^n + 7^n}$; (b) $\sum_{n=1}^{\infty} (\sqrt{n^4+1} - n^2)$; (c) $\sum_{n=1}^{\infty} \frac{n^3 3^n + n^5 5^n}{5^n + 4^n}$;

(d) $\sum_{n=1}^{\infty} \frac{\sqrt{n}+1}{n+1}$; (e) $\sum_{n=1}^{\infty} \frac{n^n}{(n+1)^{n+1}}$; (f) $\sum_{n=1}^{\infty} \frac{n^{10}}{\sqrt{(2n)!}}$;

(g) $\sum_{n=1}^{\infty} \frac{\sin^2 n}{n^2}$.

The comparison test and d'Alembert's test provide us with very powerful tools for testing for convergence. There is one further test for convergence of positive terms which we wish to include. This is the integral test. Strictly speaking it ought to be delayed until after the sections on functions, but it seems a pity to separate it from the other tests and so we include it at this point with a few preliminary definitions and the introduction of some notation.

Let $a$ be any real number; then the set of all real numbers $x$ such that $x \geqslant a$ is denoted by $[a, \infty)$ and it is called an interval. Thus the interval $[2, \infty)$ is the set of all real numbers $x$ such that $x \geqslant 2$.

DEFINITION 3.3.1 A function $f$ is said to be **decreasing** on $[a, \infty)$ if $f(x_1) \geqslant f(x_2)$ for all $x_1, x_2$ such that $a \leqslant x_1 \leqslant x_2$.

The reader should try sketching the graphs of some functions with the above property. It will then be transparently obvious why the term decreasing is used to describe the function.

Now let us suppose that $f(x)$ is defined for all $x \in [1, \infty)$ and that $f$ is positive, decreasing and continuous on $[1, \infty)$. (*Note:* For an explanation of the term continuous the reader is asked to consult Chapter 4.) Write

$$I_n = \int_1^n f(x) \, dx \qquad (n \geqslant 1), \tag{1}$$

$$s_n = f(1) + f(2) + \ldots + f(n) = \sum_{k=1}^{n} f(k). \tag{2}$$

Then it can be shown that $(s_n - I_n)$ tends to a finite limit as $n \to \infty$.

It may prove illuminating and instructive to begin by drawing some diagrams and using geometric reasoning. Before drawing any such diagrams, we must, of course, remember that $f$ is assumed to have certain properties, viz. the function $f$ is continuous, decreasing and positive (see Fig. 3.1).

We draw vertical lines at $x = 1, x = 2, x = 3, x = 4$, etc., and also horizontal lines to give the rectangles below the curve which are shown in Fig. 3.1. The rectangle on the extreme left-hand side has height $f(2)$ and width 1. Its area is, therefore, $f(2)$. The next rectangle has height $f(3)$ and width 1. Its area is therefore $f(3)$. The last rectangle, which appears at the far right-hand side of the paper, has height $f(n)$ and width 1. Its area is thus $f(n)$. The sum of the

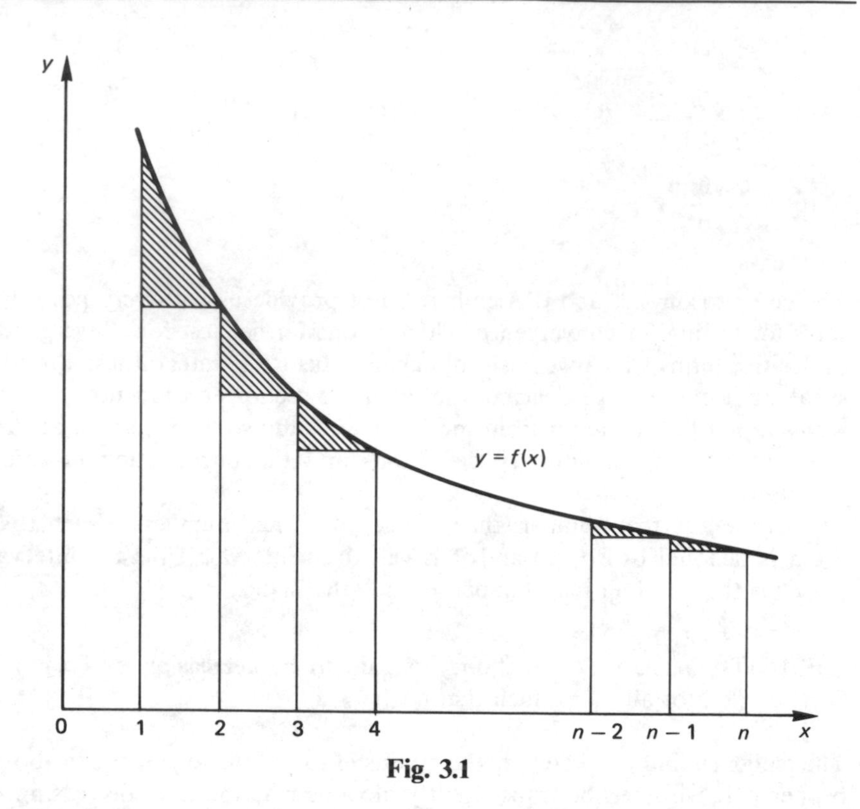


Fig. 3.1

areas of all these rectangles is $f(2)+f(3)+f(4)+\ldots+f(n)$. From (2), it follows that the sum of areas of all the rectangles below the curve between $x=1$ and $x=n$ is $s_n-f(1)$. Now the area under the curve between $x=1$ and $x=n$ is $\int_1^n f(x)\,dx$, i.e. the area under the curve is $I_n$ (see equation (1)). It follows that the difference $I_n-[s_n-f(1)]$ measures the sum of the areas of the $(n-1)$ shaded regions between the curve $y=f(x)$ and the rectangles. As $n$ increases, the number of shaded regions increases and the sum of their areas increases, i.e. $I_n-s_n+f(1)$ increases as $n$ increases. Hence $I_n-s_n$ increases as $n$ increases, since $f(1)$ is a constant.

Now we could move each of these shaded areas sideways to the left (see Fig. 3.2) and they would fit neatly into the rectangle $ABCD$ without any overlap, since $f(x)$ decreases as $x$ increases. The total area of these regions together cannot exceed the area of the rectangle $ABCD$. Now this rectangle has height $f(1)$ and width 1 and its area is therefore $f(1)$.

Hence $$I_n-s_n+f(1)\leqslant f(1)$$

i.e. $$I_n-s_n\leqslant 0.$$

and the sequence $(I_n-s_n)$ is bounded above. Since the sequence is also increasing, it must converge, i.e. $I_n-s_n$ tends to a finite limit as $n\to\infty$.

If we wish to give a proof which uses only symbols and avoids geometric reasoning, then we simply need to show that $(I_n-s_n)$ is increasing and bounded above. Let us first write

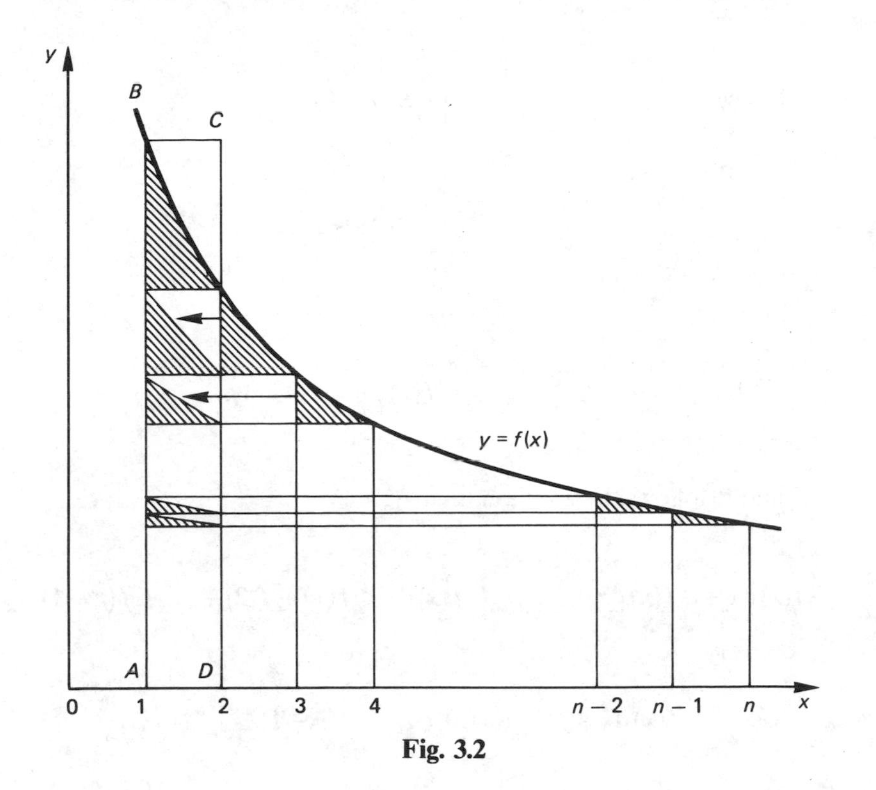


**Fig. 3.2**

$$c_n = I_n - s_n. \tag{3}$$

(a) $(I_n - s_n)$ *is increasing* For all $n \geqslant 1$,

$$\begin{aligned}
c_{n+1} - c_n &= (I_{n+1} - s_{n+1}) - (I_n - s_n) \\
&= (I_{n+1} - I_n) - (s_{n+1} - s_n) \\
&= \left( \int_1^{n+1} f(x)\,dx - \int_1^n f(x)\,dx \right) - f(n+1) \\
&= \int_n^{n+1} f(x)\,dx - f(n+1) \geqslant 0,
\end{aligned}$$

since $f(x) \geqslant f(n+1)$ for $n \leqslant x \leqslant n+1$. Thus $c_{n+1} - c_n \geqslant 0$, i.e. $c_{n+1} \geqslant c_n$ and the sequence is increasing.

(b) $(I_n - s_n)$ *is bounded above* Since $f$ is decreasing we know that for $i \leqslant x \leqslant i+1$ $(i = 1, 2, 3, 4, \dots)$ $f(x) \leqslant f(i)$, and therefore

$$\int_i^{i+1} f(x)\,dx \leqslant \int_i^{i+1} f(i)\,dx = f(i) \int_i^{i+1} dx = f(i).$$

Hence
$$\int_1^2 f(x)\,dx \leqslant f(1),$$
$$\int_2^3 f(x)\,dx \leqslant f(2),$$
$$\cdots\cdots\cdots\cdots\cdots\cdots$$
$$\int_{n-1}^n f(x)\,dx \leqslant f(n-1),$$

and it follows, by addition, that

$$\int_1^2 f(x)\,dx + \int_2^3 f(x)\,dx + \ldots + \int_{n-1}^n f(x)\,dx \leqslant f(1) + f(2) + \ldots + f(n-1),$$

i.e. $$\int_1^n f(x)\,dx \leqslant f(1) + f(2) + \ldots + f(n-1).$$

Since $f(x) \geqslant 0$ for all $x \geqslant 1$, we see that

$$\begin{aligned}\int_1^n f(x)\,dx &\leqslant f(1) + f(2) + \ldots + f(n-1)\\ &\leqslant f(1) + f(2) + \ldots + f(n-1) + f(n)\\ &= s_n;\end{aligned}$$

i.e. $$I_n \leqslant s_n$$

and therefore $I_n - s_n \leqslant 0 \ (n \geqslant 1)$.

Since $(I_n - s_n)$ is increasing and bounded above, $I_n - s_n$ tends to a finite limit as $n \to \infty$.

In order to keep a record, for future reference, we incorporate these results in a theorem.

THEOREM 3.3.4 Let the function $f$ be continuous, decreasing and positive on $[1, \infty)$. Write

$$s_n = f(1) + f(2) + \ldots + f(n) \qquad (n \geqslant 1),$$

$$I_n = \int_1^n f(x)\,dx \qquad (n \geqslant 1).$$

Then $S_n - I_n$ tends to a finite limit as $n \to \infty$.

The integral test is now a simple deduction from this theorem.

THEOREM 3.3.5 (Integral Test) Let the function $f$ be continuous, decreasing and positive on $[1, \infty)$. Then

(a) $\sum_{n=1}^{\infty} f(n)$ converges, if $\int_1^n f(x)\,dx$ tends to a finite limit as $n \to \infty$,
(b) $\sum_{n=1}^{\infty} f(n)$ diverges, if $\int_1^n f(x)\,dx$ does not tend to a finite limit as $n \to \infty$.

*Proof* We use the standard notation

$$s_n = f(1) + f(2) + \ldots + f(n) \qquad (1)$$

$$I_n = \int_1^n f(x)\,dx \qquad (2)$$

and we write

$$c_n = s_n - I_n. \qquad (3)$$

Clearly $c_n$ tends to a finite limit as $n \to \infty$, by Theorem 3.3.4.

Since $f(x) > 0$ for $x \geqslant 1$, we see that $s_n$ and $I_n$ both increase as $n$ increases. Thus either

(a) $(I_n)$ is bounded above and $I_n$ tends to a finite limit as $n \to \infty$, or
(b) $(I_n)$ is not bounded above and $I_n \to \infty$ as $n \to \infty$.

(a) If $I_n$ tends to a finite limit as $n \to \infty$, then $s_n = I_n + c_n$ must also tend to a finite limit as $n \to \infty$ (because $c_n$ tends to a finite limit as $n \to \infty$). Hence $\sum_{n=1}^{\infty} f(n)$ converges, which proves (a).
(b) If $I_n \to \infty$, then $s_n = c_n + I_n \to \infty$ as $n \to \infty$ and therefore $\sum_{n=1}^{\infty} f(n)$ diverges, which proves (b).

It is, of course, not vital to start at $x = 1$; we could start at $x = N_0$ where $N_0$ is any given positive integer, provided all the conditions are satisfied on $[N_0, \infty)$. This gives

COROLLARY Let the function $f$ be continuous, decreasing and positive on $[N_0, \infty)$ where $N_0$ is some given positive integer. Then

(a) $\sum_{n=N_0}^{\infty} f(n)$ converges if $\int_{N_0}^n f(x)\,dx$ tends to a finite limit as $n \to \infty$,

(b) $\sum_{n=N_0}^{\infty} f(n)$ diverges if $\int_{N_0}^{n} f(x)\,dx$ does not tend to a finite limit as $n \to \infty$.

So far all our integrals have been of the form $\int_{N_0}^{n} f(x)\,dx$ where $N_0, n$ are positive integers. With the given conditions on $f$ it is possible to prove that $\int_{N_0}^{X} f(x)\,dx$ tends to a finite limit as $X \to \infty$ if and only if $\int_{N_0}^{n} f(x)\,dx$ tends to a finite limit as $n \to \infty$. We say that $\int_{N_0}^{\infty} f(x)\,dx$ converges if and only if $\int_{N_0}^{X} f(x)\,dx$ tends to a finite limit as $X \to \infty$. The integral test is, therefore, sometimes recorded in the following form.

THEOREM 3.3.5(a) (Integral Test) Let the function $f$ be continuous, decreasing and positive on $[N_0, \infty)$, where $N_0$ is some given positive integer. Then $\sum_{n=N_0}^{\infty} f(n)$ and $\int_{N_0}^{\infty} f(x)\,dx$ either both converge or both diverge.

On the whole, the introduction of $\int_{N_0}^{\infty} f(x)\,dx$ and the idea of convergence of integrals seems over-elaborate at this stage, since it is quite sufficient to consider only $\lim_{n\to\infty} \int_{N_0}^{n} f(x)\,dx$. We will therefore use Theorem 3.3.5 and its corollary in examples. The reader must, however, be aware that many authors will use the second form (Theorem 3.3.5(a)) and must be prepared to meet it in books on analysis.

Our first example, using the integral test, settles a problem which arose earlier in the chapter, viz. does the series $\sum_{n=1}^{\infty} 1/n^{\alpha}$ converge if $1 < \alpha < 2$? Fortunately the answer is yes. This means that we now know that

$$\boxed{\sum_{n=1}^{\infty} \frac{1}{n^{\alpha}} \text{ converges if } \alpha > 1,}$$

and

$$\boxed{\sum_{n=1}^{\infty} \frac{1}{n^{\alpha}} \text{ diverges if } \alpha \leqslant 1.}$$

Let us prove the result which closes the gap left earlier in the chapter.

---

*Examples 3.3.5*

1. Let $f(x) = \dfrac{1}{x^{\alpha}}$ $(x \geqslant 1)$, where $\alpha > 0$. Then $f$ is continuous, decreasing and positive on $[1, \infty)$. Using the standard notation we write

$$I_n = \int_1^n f(x)\,dx \qquad (n = 1, 2, 3, 4, \ldots).$$

Then

$$I_n = \int_1^n \frac{1}{x^{\alpha}}\,dx = \begin{cases} \left[-\dfrac{1}{(\alpha-1)}\dfrac{1}{x^{\alpha-1}}\right]_1^n & (\alpha \neq 1), \\ \Big[\log x\Big]_1^n & (\alpha = 1). \end{cases}$$

Now if $\alpha > 1$, $$I_n = \frac{1}{\alpha-1}\left(1 - \frac{1}{n^{\alpha-1}}\right) \to \frac{1}{\alpha-1}$$

as $n \to \infty$, and so $\sum_{n=1}^{\infty} 1/n^{\alpha}$ converges by the integral test for $\alpha > 1$.
However, if $\alpha = 1$, $I_n = \int_1^n 1/x^{\alpha}\,dx = \log n$ and $I_n \to \infty$ as $n \to \infty$. Thus $\sum_{n=1}^{\infty} 1/n^{\alpha}$ diverges if $\alpha = 1$.
If $0 < \alpha < 1$, then $I_n \to \infty$ as $n \to \infty$, and again $\sum_{n=1}^{\infty} 1/n^{\alpha}$ diverges.

The next example illustrates the fact that it is sometimes necessary to start at a point other than 1.

2. We use the integral test to decide whether $\sum_{n=2}^{\infty} 1/[n \log n]$ converges. To do this we use

$$f(x) = \frac{1}{x \log x} \qquad (x \geqslant 2).$$

Then $f$ is continuous, decreasing and positive on $[2, \infty)$. Moreover,

$$I_n = \int_2^n f(x)\,dx = \int_2^n \frac{1}{x \log x}\,dx = [\log(\log x)]_2^n$$

$$= \log(\log n) - \log(\log 2).$$

Hence $I_n \to \infty$ as $n \to \infty$, and so $\sum_{n=2}^{\infty} 1/[n \log n]$ diverges.

If Theorem 3.3.4 is applied with the $f(x) = \frac{1}{x}$, we derive some rather interesting results.

3. For if $f(x) = \frac{1}{x}$ for $x \geqslant 1$, then $f$ is decreasing, continuous and positive on $[1, \infty)$. With the standard notation, we have

$$I_n = \int_1^n f(x)\,dx = \int_1^n \frac{1}{x}\,dx = [\log x]_1^n = \log n \tag{1}$$

and $$s_n = \sum_{k=1}^{n} \frac{1}{k}.$$

$$= 1 + \tfrac{1}{2} + \tfrac{1}{3} + \tfrac{1}{4} + \ldots + \tfrac{1}{n}.$$

Write $$\gamma_n = s_n - I_n. \tag{2}$$

Then, from Theorem 3.3.4, $\gamma_n$ tends to a finite limit $\gamma$ as $n \to \infty$. This number $\gamma$ is called Euler's constant. Its value has been computed to over 200 decimal places. It is 0.577 215 664 901 53 correct to fourteen decimal places.
It follows from (1) and (2) that

$$s_n = I_n + \gamma_n = \log n + \gamma_n,$$

where $\gamma_n \to \gamma$ as $n \to \infty$.

i.e. $$1+\tfrac{1}{2}+\tfrac{1}{3}+\ldots+\tfrac{1}{n}=\log n+\gamma_n. \qquad (3)$$

This relation can be used to sum certain series.

*Note:* Formula (3) shows that the sum $1+\frac{1}{2}+\frac{1}{3}+\frac{1}{4}+\ldots+\frac{1}{n}$ behaves like $\log n+\gamma$ for large $n$. We already know that $s_n\to\infty$ as $n\to\infty$, but we now have a better idea of how rapidly $s_n$ grows as $n$ increases.

4. Consider the series

$$1-\frac{1}{2}+\frac{1}{3}-\frac{1}{4}+\frac{1}{5}-\frac{1}{6}+\ldots=\sum_{n=1}^{\infty}(-1)^{n+1}\frac{1}{n}.$$

Let $t_n$ be the sum of the first $n$ terms. Because the signs alternate plus, minus, plus, minus, etc., it is simpler to first consider an even number of terms—say the first $2n$ terms. The sum of these first $2n$ terms is

$$\begin{aligned} t_{2n} &= 1-\frac{1}{2}+\frac{1}{3}-\frac{1}{4}+\ldots+\frac{1}{2n-1}-\frac{1}{2n} \\ &= \left(1+\frac{1}{2}+\frac{1}{3}+\frac{1}{4}+\ldots+\frac{1}{2n-1}+\frac{1}{2n}\right) \\ &\quad -2\left(\frac{1}{2}+\frac{1}{4}+\frac{1}{6}+\frac{1}{8}+\ldots+\frac{1}{2n}\right) \\ &= \left(1+\frac{1}{2}+\frac{1}{3}+\frac{1}{4}+\ldots+\frac{1}{2n-1}+\frac{1}{2n}\right) \\ &\quad -\left(1+\frac{1}{2}+\frac{1}{3}+\frac{1}{4}+\ldots+\frac{1}{n}\right) \\ &= \log 2n+\gamma_{2n}-[\log n+\gamma_n] \end{aligned}$$

from equation (3) above. Hence

$$\begin{aligned} t_{2n} &= \log 2n-\log n+\gamma_{2n}-\gamma_n \\ &= \log 2+\gamma_{2n}-\gamma_n \end{aligned}$$

and $$t_{2n}\to\log 2+\gamma-\gamma=\log 2$$

as $n\to\infty$.

If we add an odd number of terms (say the first $2n+1$ terms) we have

$$\begin{aligned} t_{2n+1} &= 1-\frac{1}{2}+\frac{1}{3}-\frac{1}{4}+\ldots+\frac{1}{2n-1}-\frac{1}{2n}+\frac{1}{2n+1} \\ &= t_{2n}+\frac{1}{2n+1}, \end{aligned}$$

and therefore $$t_{2n+1}\to\log 2+0=\log 2$$

as $n \to \infty$. Since $t_{2n} \to \log 2$ and $t_{2n+1} \to \log 2$ as $n \to \infty$, we see that $t_n \to \log 2$ as $n \to \infty$. The given series, therefore, converges and its sum is $\log 2$,

i.e. $$1 - \frac{1}{2} + \frac{1}{3} - \frac{1}{4} + \ldots = \log 2.$$

5. Suppose we now take the above series and rearrange its terms as

$$1 + \frac{1}{3} + \frac{1}{5} + \frac{1}{7} - \frac{1}{2} + \frac{1}{9} + \frac{1}{11} + \frac{1}{13} + \frac{1}{15} - \frac{1}{4} + \ldots,$$

i.e. we take four positive terms, then one negative, then four positive terms etc. Will this rearranged series still have sum $\log 2$? Well, let us see. Let $r_n$ be the sum of the first $n$ terms. Obviously it will be easier if we first see what happens if we take the terms in bundles of five (four positive and one negative term in each bundle). Let us take $n$ bundles of 5, i.e. $5n$ terms in all. The sum $r_{5n}$ of these first $5n$ terms is given by

$$r_{5n} = 1 + \frac{1}{3} + \frac{1}{5} + \frac{1}{7} - \frac{1}{2} + \frac{1}{9} + \frac{1}{11} + \frac{1}{13} + \frac{1}{15} - \frac{1}{4} + \ldots$$

$$+ \frac{1}{8n-7} + \frac{1}{8n-5} + \frac{1}{8n-3} + \frac{1}{8n-1} - \frac{1}{2n}$$

$$= \left(1 + \frac{1}{2} + \frac{1}{3} + \frac{1}{4} + \ldots + \frac{1}{8n}\right) - \left(\frac{1}{2} + \frac{1}{4} + \frac{1}{6} + \frac{1}{8} + \ldots + \frac{1}{8n}\right)$$

$$- \left(\frac{1}{2} + \frac{1}{4} + \frac{1}{6} + \frac{1}{8} + \ldots + \frac{1}{2n}\right).$$

(*Note:* Many students frequently find it difficult at first to write down suitable expressions for the terms in the $n$th bracket. The following reasoning may help. The initial terms in each of the first three brackets are $\frac{1}{1}$, $\frac{1}{9}$ and $\frac{1}{17}$ respectively. The denominators increase by 8, and therefore the positive terms have denominators of the form $8n +$ something. Now the adjustment is easy. Similarly, the denominators of the negative term go up by 2, and they therefore contain $2n +$ something.)

The first two brackets taken together (cancelling out terms as appropriate) give the positive terms and the third bracket contains the negative terms. Hence, from (3),

$$r_{5n} = \left(1 + \frac{1}{2} + \frac{1}{3} + \frac{1}{4} + \ldots + \frac{1}{8n}\right) - \frac{1}{2}\left(1 + \frac{1}{2} + \frac{1}{3} + \ldots + \frac{1}{4n}\right)$$

$$- \frac{1}{2}\left(1 + \frac{1}{2} + \frac{1}{3} + \frac{1}{4} + \ldots + \frac{1}{n}\right)$$

$$= \log 8n + \gamma_{8n} - \tfrac{1}{2}[\log 4n + \gamma_{4n}] - \tfrac{1}{2}[\log n + \gamma_n]$$

$$= \log 8n - \tfrac{1}{2}\log 4n - \tfrac{1}{2}\log n + \gamma_{8n} - \tfrac{1}{2}\gamma_{4n} - \tfrac{1}{2}\gamma_n$$

$$= \tfrac{1}{2}\log\left(\frac{8n \cdot 8n}{4n \cdot n}\right) + \gamma_{8n} - \tfrac{1}{2}\gamma_{4n} - \tfrac{1}{2}\gamma_n$$

$$= \tfrac{1}{2}\log 16 + \gamma_{8n} - \tfrac{1}{2}\gamma_{4n} - \tfrac{1}{2}\gamma_n$$

$$\to \tfrac{1}{2}\log 16 + \gamma - \tfrac{1}{2}\gamma - \tfrac{1}{2}\gamma = \tfrac{1}{2}\log 16 = \log 4.$$

as $n \to \infty$. Thus $r_{5n} \to \log 4$ as $n \to \infty$. Now, we know that the positive integers are not all divisible by 5. Thus if we take an arbitrary (finite) number of terms and go along the series collecting these terms in bundles of five, then there may be some terms left on their own at the end. This remainder could comprise 0, 1, 2, 3 or 4 extra terms. We know $r_{5n} \to \log 4$ as $n \to \infty$. This takes care of the case in which there are no remaining terms. We must, therefore, also consider what happens to $r_{5n+1}$, $r_{5n+2}$, $r_{5n+3}$, $r_{5n+4}$. Now as $n \to \infty$,

$$r_{5n+1} = r_{5n} + \frac{1}{8n+1} \to \log 4 + 0 = \log 4,$$

$$r_{5n+2} = r_{5n} + \frac{1}{8n+1} + \frac{1}{8n+3} \to \log 4,$$

$$r_{5n+3} = r_{5n} + \frac{1}{8n+1} + \frac{1}{8n+3} + \frac{1}{8n+5} \to \log 4,$$

$$r_{5n+4} = r_{5n} + \frac{1}{8n+1} + \frac{1}{8n+3} + \frac{1}{8n+5} + \frac{1}{8n+7} \to \log 4.$$

Gathering all the results together, we see that $r_n \to \log 4$ as $n \to \infty$. Thus the given series converges and its sum is $\log 4$.

Now the series in the above example is only a rearrangement of the series $1 - \frac{1}{2} + \frac{1}{3} - \frac{1}{4} + \frac{1}{5} - \frac{1}{6} + \ldots$, which has sum $\log 2$. So we have, clearly, altered the sum of the series by rearranging the order of its terms. Does this always happen? For the answer to this question, the reader will have to be patient and wait a little while!

---

The reader may also wonder how many different sums can be obtained by rearranging such a series. The answer is that the series

$$1 - \tfrac{1}{2} + \tfrac{1}{3} - \tfrac{1}{4} + \tfrac{1}{5} - \tfrac{1}{6} + \ldots$$

can be rearranged in such a way as to have any required sum, i.e. if the reader chooses a favourite real number $x$, then there is a rearrangement whose sum is $x$. Except for particular values of $x$ it is not possible to write down a nice, neat formula specifying which term is in the $n$th position of any rearrangement with sum $x$. This does not matter. The important thing is that such rearrangements exist. (*Note:* The reader curious to know a little more about this should consult the appendix at the end of this chapter, which contains the outline of one possible method for constructing such rearrangements.)

In previous sections we have seen that the series

$$1-\tfrac{1}{2}+\tfrac{1}{3}-\tfrac{1}{4}+\tfrac{1}{5}-\tfrac{1}{6}+\tfrac{1}{7}-\tfrac{1}{8}+\dots$$

converges—its sum being $\log 2$. This is a series in which the signs alternate, plus, minus, plus, minus, etc. It is of the form $\sum_{n=1}^{\infty}(-1)^{n+1}a_n$, where $a_n=\frac{1}{n}$ $(n=1,2,3,\dots)$. In this case

(a) $a_n>0$ for all $n$,
(b) $a_{n+1}\leqslant a_n$, i.e. the sequence $(a_n)$ is decreasing,
(c) $a_n\to 0$ as $n\to\infty$.

Are these properties of $a_n$ sufficient to guarantee that every such series of the form $\sum_{n=1}^{\infty}(-1)^{n+1}a_n$ converges? The answer is yes. The result, which is stated below, is normally called either the alternating series test or the alternating signs test. Strictly speaking it should, probably, be called the test for series whose terms have alternating signs, but this would be rather a mouthful!

THEOREM 3.3.6 (Alternating Series Test) Let $(a_n)_{n=1}^{\infty}$ be a decreasing sequence of positive real numbers such that $a_n\to 0$ as $n\to\infty$. Then the series $\sum_{n=1}^{\infty}(-1)^{n+1}a_n$ converges.

*Proof* Since all the previous tests (except Theorem 3.3.1) are for series with positive terms, they are not applicable. We must therefore go back to the definition and consider the behaviour of the partial sums $s_n$, where

$$s_n=a_1-a_2+a_3-a_4+\dots+(-1)^{n+1}a_n.$$

Initially, it is probably easiest to consider the sum of an even number of terms. The sum of the first $2n$ terms is given by

$$s_{2n}=a_1-a_2+a_3-a_4+\dots+a_{2n-1}-a_{2n}.$$

Hence
$$s_{2n+2}-s_{2n}=a_{2n+1}-a_{2n+2}\geqslant 0,$$

since $(a_n)$ is decreasing and, therefore, $a_{2n+1}\geqslant a_{2n+2}$. Thus

$$s_{2n+2}\geqslant s_{2n}$$

and the sequence $(s_{2n})$ is increasing. Moreover,

$$\begin{aligned}s_{2n}&=a_1-a_2+a_3-a_4+a_5-\dots-a_{2n-2}+a_{2n-1}-a_{2n}\\&=a_1-(a_2-a_3)-(a_4-a_5)-\dots-(a_{2n-2}-a_{2n-1})-a_{2n}\\&\leqslant a_1\end{aligned}$$

because $a_2 \geqslant a_3 \geqslant a_4 \geqslant \ldots \geqslant a_{2n-2} \geqslant a_{2n-1}$ and $a_{2n} > 0$. Thus $s_{2n} \leqslant a_1$ for all $n$ and the sequence $(s_{2n})$ is bounded above. Since it is also increasing, it must converge to some number $s$, i.e. $s_{2n} \to s$ as $n \to \infty$. Of course, we must also investigate what happens if we take an odd number of terms. The sum of the first $2n+1$ terms is $s_{2n+1}$, where

$$\begin{aligned} s_{2n+1} &= a_1 - a_2 + a_3 - a_4 + \ldots + a_{2n-1} - a_{2n} + a_{2n+1} \\ &= s_{2n} + a_{2n+1}. \end{aligned}$$

As $n \to \infty$, $a_{2n+1} \to 0$ and $s_{2n} \to s$. It follows that $s_{2n+1} \to s$ as $n \to \infty$. Since $s_{2n} \to s$ and $s_{2n+1} \to s$ as $n \to \infty$, we see that $s_n \to s$ as $n \to \infty$ and the series, therefore, converges (and its sum is $s$).

If, however, $a_n \nrightarrow 0$ as $n \to \infty$, then $(-1)^{n+1} a_n \nrightarrow 0$ as $n \to \infty$ and the series $\sum_{n=1}^{\infty} (-1)^{n+1} a_n$ diverges by Theorem 3.3.1.

---

*Examples 3.3.6*

1. Let $a_n = \sin(1/n)$ $(n = 1, 2, 3, \ldots)$. For $n \geqslant 1$,

$$0 < \frac{1}{n+1} < \frac{1}{n} \leqslant 1 < \frac{\pi}{2},$$

and, therefore, $\quad a_n = \sin\left(\frac{1}{n}\right) > 0,$

$$a_{n+1} = \sin\left(\frac{1}{n+1}\right) < \sin\left(\frac{1}{n}\right) = a_n,$$

i.e. $(a_n)$ is a decreasing sequence of positive real numbers. Furthermore, $a_n = \sin(1/n) \to 0$ as $n \to \infty$. By the alternating series test,

$$\sum_{n=1}^{\infty} (-1)^{n+1} \sin(1/n)$$

converges.

But what happens to this series if we dispense with the factor $(-1)^{n+1}$? Using the limit form of the comparison test (with $a_n = \sin(1/n)$, $b_n = 1/n$), we can show that $\sum_{n=1}^{\infty} \sin(1/n)$ diverges. The factors $(-1)^{n+1}$ therefore made all the difference in this case.

2. Let $a_n = \cos(1/n)$. As $n \to \infty$, $\cos(1/n) \to 1$. Hence $(-1)^{n+1} \cos(1/n) \nrightarrow 0$ as $n \to \infty$ and so $\sum_{n=1}^{\infty} (-1)^{n+1} \cos(1/n)$ diverges by Theorem 3.3.1.

3. Let $a_n = \sin^2(1/n)$. As in Example 1, it is easy to show that $(a_n)_{n=1}^{\infty}$ is a decreasing sequence of positive real numbers, such that $a_n \to 0$ as $n \to \infty$. By the alternating series test $\sum_{n=1}^{\infty} (-1)^{n+1} \sin^2(1/n)$ converges.

Again, let us see what happens if we drop the factors $(-1)^{n+1}$. Using the limit form of the comparison test (with $a_n = \sin^2(1/n)$, $b_n = 1/n^2$), we can show that $\sum_{n=1}^{\infty} \sin^2(1/n)$ also converges. (This fact will later give rise to an alternative way of dealing with the question of the convergence of $\sum_{n=1}^{\infty}(-1)^{n+1}\sin^2(1/n)$.) In this case, therefore, we still have convergence even when we omit the factors $(-1)^{n+1}$. This contrasts with the series in Example 1.

---

In stating the result, we used $(-1)^{n+1}$ in preference to $(-1)^n$, as the former gives the first term of the series as $a_1$ (rather than $-a_1$). However, it really makes no difference to the conclusion if $(-1)^n$ is used in place of $(-1)^{n+1}$. For if we write

$$s_n = a_1 - a_2 + \ldots + (-1)^{n+1}a_n = \sum_{k=1}^{n}(-1)^{k+1}a_k,$$

$$t_n = -a_1 + a_2 + \ldots + (-1)^n a_n = \sum_{k=1}^{n}(-1)^k a_k,$$

then $t_n = -s_n$ and $t_n$ tends to a finite limit as $n \to \infty$ if and only if $s_n$ tends to a finite limit as $n \to \infty$. Thus if $(a_n)$ satisfies all the conditions of Theorem 3.3.6, then $\sum_{n=1}^{\infty}(-1)^n a_n$ is convergent.

## *EXERCISES 3.3.3*

**1** Prove the following series converge:

(a) $\sum_{n=2}^{\infty}(-1)^n \frac{1}{\log n}$; (b) $\sum_{n=1}^{\infty}(-1)^{n+1}\frac{1}{\sqrt{n}}$;

(c) $\sum_{n=1}^{\infty}(-1)^{n+1}(\sqrt{n+1}-\sqrt{n})$.

**2** Prove the following series diverges:

$$\sum_{n=1}^{\infty}(-1)^{n+1}\cosh\left(\frac{1}{n}\right).$$

**3** Decide whether the following series converge or diverge. Give reasons to justify your answers. (The reader must be prepared to use all the tests in section 3.3.)

(a) $\sum_{n=1}^{\infty}\sin\left(\frac{(n^2+1)\pi}{n}\right)$; (b) $\sum_{n=1}^{\infty}(\sqrt{n^3+3}-\sqrt{n^3+1})$;

(c) $\sum_{n=1}^{\infty}(-1)^{n+1}\cos^2\frac{1}{n}$; (d) $\sum_{n=2}^{\infty}\frac{1}{n(\log n)^3}$;

(e) $\sum_{n=1}^{\infty}\frac{n^7 2^n + n^2 3^n}{n^9 + n^2 3^n}$; (f) $\sum_{n=1}^{\infty}\frac{3^n(n!)^3}{(3n)!}$.

## 3.4 ABSOLUTE CONVERGENCE, MULTIPLICATION OF SERIES, REARRANGEMENT OF SERIES

In the previous section we saw that the series $\sum_{n=1}^{\infty}(-1)^{n+1}\sin^2(1/n)$ is convergent and the related series $\sum_{n=1}^{\infty}\sin^2(1/n)$ is also convergent. That is, $\sum_{n=1}^{\infty}|(-1)^{n+1}\sin^2(1/n)|$ is convergent. Such a series is said to be absolutely convergent.

DEFINITION 3.4.1 Let $\sum_{n=1}^{\infty}a_n$ be a series of real numbers. If $\sum_{n=1}^{\infty}|a_n|$ is convergent then $\sum_{n=1}^{\infty}a_n$ is said to be **absolutely convergent**.

The phrase 'absolute convergence' immediately suggests convergence plus something more. We would, therefore, expect that an absolutely convergent series is certainly convergent. Fortunately, this is indeed the case.

THEOREM 3.4.1 Every absolutely convergent series of real numbers is also convergent.

*Proof* Let $\sum_{n=1}^{\infty}a_n$ be an absolutely convergent series of real numbers, i.e. it is a series such that $\sum_{n=1}^{\infty}|a_n|$ converges. Now some of the terms of $\sum_{n=1}^{\infty}a_n$ may be positive and some may be negative. We therefore separate out the positive and negative terms in the following way.

Let $$b_n=\begin{cases}a_n & \text{if } a_n\geqslant 0,\\ 0 & \text{if } a_n<0,\end{cases}\qquad c_n=\begin{cases}0 & \text{if } a_n\geqslant 0,\\ -a_n & \text{if } a_n<0.\end{cases}$$

Then $b_n\geqslant 0$, $c_n\geqslant 0$ for all $n$, and

$$a_n=b_n-c_n. \tag{1}$$

Write

$$s_n=|a_1|+|a_2|+\ldots+|a_n|, \tag{2}$$

$$t_n=a_1+a_2+\ldots+a_n,$$

$$r_n=b_1+b_2+\ldots+b_n, \tag{3}$$

$$u_n=c_1+c_2+\ldots+c_n. \tag{4}$$

Now we are given that $\sum_{n=1}^{\infty}|a_n|$ converges and so $s_n$ tends to a finite limit $s$ as $n\to\infty$. Moreover, $|a_n|\geqslant 0$ for all $n$ (by definition of the modulus) and so the sequence $(s_n)$ is increasing. Thus

$$s_n\leqslant s$$

for all $n$. Now, from the definition of $b_n$, we see that

$$r_n=b_1+b_2+\ldots+b_n\leqslant|a_1|+|a_2|+\ldots+|a_n|=s_n\leqslant s$$

for all $n$,

i.e.
$$r_n \leqslant s$$

for all $n$. But $b_n \geqslant 0$ for all $n$ and therefore the sequence $(r_n)$ is increasing. Since it is bounded above by $s$, $(r_n)$ must converge to some limit $r$ as $n \to \infty$. Similarly, $(u_n)$ converges to some limit $u$ as $n \to \infty$. Now from (1), (2), (3) and (4)

$$t_n = r_n - u_n \to r - u$$

as $n \to \infty$. Since $t_n$ tends to a finite limit as $n \to \infty$, $\sum_{n=1}^{\infty} a_n$ converges as required.

We have, therefore, proved that an absolutely convergent series of real numbers is also convergent. The converse, however, is not true. There are series which are convergent, but not absolutely convergent. For example, the series $\sum_{n=1}^{\infty} (-1)^{n+1} 1/n$ is convergent. This is just the familiar series

$$1 - \tfrac{1}{2} + \tfrac{1}{3} - \tfrac{1}{4} + \tfrac{1}{5} - \tfrac{1}{6} + \dots.$$

However, it is not absolutely convergent, since $\sum_{n=1}^{\infty} 1/n$ diverges. Such a series is called a conditionally convergent series.

DEFINITION 3.4.2 A series of real numbers which is convergent, but not absolutely convergent, is called a **conditionally convergent series.**

*Example 3.4.1*

In the previous section we showed that $\sum_{n=1}^{\infty} (-1)^{n+1} \sin^2(1/n)$ is convergent by using the alternating series test. It would, however, probably be easier to prove absolute convergence, which of course implies convergence. We notice that

$$|(-1)^{n+1} \sin^2(1/n)| = \sin^2(1/n).$$

Now write $\quad a_n = \sin^2(1/n), \qquad b_n = 1/n^2 \qquad (n = 1, 2, 3, \dots).$

Then $a_n > 0$, $b_n > 0$ and

$$\frac{a_n}{b_n} = \left(\frac{\sin(1/n)}{(1/n)}\right)\left(\frac{\sin(1/n)}{(1/n)}\right) \to 1.1 = 1 \text{ as } n \to \infty.$$

Now $\sum_{n=1}^{\infty} 1/n^2$ converges, and therefore, by the limit form of the comparison, test $\sum_{n=1}^{\infty} \sin^2(1/n)$ is convergent, i.e. $\sum_{n=1}^{\infty} |(-1)^{n+1} \sin^2(1/n)|$ is convergent. Thus $\sum_{n=1}^{\infty} (-1)^{n+1} \sin^2(1/n)$ is absolutely convergent and therefore convergent.

We have introduced the notion of absolute convergence. But what have we gained? What advantages does absolute convergence bestow on us? To obtain some insight we go back to some ideas introduced in the previous section. We found that the conditionally convergent series

$$1-\tfrac{1}{2}+\tfrac{1}{3}-\tfrac{1}{4}+\tfrac{1}{5}-\tfrac{1}{6}+\dots$$

could be rearranged to give any desired sum. However, an absolutely convergent series does not have this property. In fact every rearrangement of an absolutely convergent series has the same sum. So absolutely convergent series can be rearranged (and multiplied) without changing the sum. The proof of this is quite long and rather complicated. Some readers may, therefore, wish to omit the detailed proofs (which comprise the remainder of this chapter) and may be content to make a note of the following statements.

(a) Every rearrangement of an absolutely convergent series has the same sum as the original series.
(b) Absolutely convergent series can be multiplied.

For the sake of the brave souls who refuse to be deterred by threats of long complicated proofs, we begin by showing that any rearrangement of an absolutely convergent series is also absolutely convergent. The question of whether the sum of the series is altered by the change of order of the terms is deferred for just a little while.

THEOREM 3.4.2 Let $\sum_{n=1}^{\infty} a_n$ be an absolutely convergent series. Then every rearrangement of it is also absolutely convergent. Moreover, if $\sum_{n=1}^{\infty} v_n$ is any such rearrangement, then

$$\sum_{n=1}^{\infty} |a_n| = \sum_{n=1}^{\infty} |v_n|.$$

*Proof* Let $\sum_{n=1}^{\infty} v_n$ be a rearrangement of $\sum_{n=1}^{\infty} a_n$. Write

$$s_n = |a_1| + |a_2| + \dots + |a_n|,$$
$$t_n = |v_1| + |v_2| + \dots + |v_n|.$$

Since $\sum_{n=1}^{\infty} a_n$ is absolutely convergent (i.e. $\sum_{n=1}^{\infty} |a_n|$ converges), $s_n$ tends to a finite limit $s$ as $n \to \infty$. Now all the terms $|a_n|$ are non-negative and so $(s_n)$ is an increasing sequence which converges to $s$. Thus

$$s_n \leqslant s \tag{1}$$

for all $n$. Now if $n$ is any given positive integer, there is a corresponding positive integer $M$ such that all the terms $|v_1|, |v_2|, |v_3|, \dots, |v_n|$ (and probably many more) belong to the set $\{|a_1|, |a_2|, |a_3|, \dots, |a_M|\}$ (since $\sum_{n=1}^{\infty} |v_n|$ is a rearrangement of $\sum_{n=1}^{\infty} |a_n|$. Hence

$$t_n = |v_1| + |v_2| + \ldots + |v_n| \leqslant |a_1| + |a_2| + \ldots + |a_M| = s_M \leqslant s$$

by relation (1). Thus
$$t_n \leqslant s \tag{2}$$
for all $n$. Now the terms $|v_n|$ are all non-negative. Hence $(t_n)$ is an increasing sequence. Since $(t_n)$ is also bounded above (see relation (2)), it follows that $t_n$ tends to a finite limit $t$ as $n \to \infty$, where

$$t \leqslant s \tag{3}$$

by (2). This means that $\sum_{n=1}^{\infty} |v_n|$ converges, i.e. $\sum_{n=1}^{\infty} v_n$ is absolutely convergent and $\sum_{n=1}^{\infty} |v_n| = t$, where $t \leqslant s$ by relation (3).

The inequality $t \leqslant s$ is a direct consequence of the assumption that $\sum_{n=1}^{\infty} |v_n|$ is a rearrangement of $\sum_{n=1}^{\infty} |a_n|$. However, we could equally well treat $\sum_{n=1}^{\infty} |a_n|$ as a rearrangement of the series $\sum_{n=1}^{\infty} |v_n|$. The preceding argument (with $a_n$ and $v_n$ interchanged) would then give

$$s \leqslant t \tag{4}$$

Relations (3) and (4) together give

$$s = t,$$

i.e.
$$\sum_{n=1}^{\infty} |a_n| = \sum_{n=1}^{\infty} |v_n|.$$

Now we can address the question of whether the sum of an absolutely convergent series is altered by rearrangement. Let us begin with the special case of a series whose terms are all non-negative real numbers.

THEOREM 3.4.3 Let $\sum_{n=1}^{\infty} a_n$ be an absolutely convergent series of non-negative real numbers. Then every rearrangement of this series is absolutely convergent and has the same sum.

*Proof* Let $\sum_{n=1}^{\infty} v_n$ be a rearrangement of $\sum_{n=1}^{\infty} a_n$. Since $a_n \geqslant 0$ for all $n$, it follows that $v_n \geqslant 0$ for all $n$ and

$$|a_n| = a_n, \qquad |v_n| = v_n. \tag{1}$$

In this case, therefore, absolute convergence and convergence mean the same thing. Using the previous theorem we see that $\sum_{n=1}^{\infty} v_n$ is absolutely convergent and

$$\sum_{n=1}^{\infty} |a_n| = \sum_{n=1}^{\infty} |v_n|.$$

In view of relation (1) this gives

$$\sum_{n=1}^{\infty} a_n = \sum_{n=1}^{\infty} v_n$$

and the proof is complete.

But what happens in the general case when the series contains both positive and negative terms? In this case, we separate out the positive and negative terms, in much the same way as in the proof of Theorem 3.4.1.

THEOREM 3.4.4 Let $\sum_{n=1}^{\infty} a_n$ be an absolutely convergent series of real numbers. Then every rearrangement of this series is also absolutely convergent and has the same sum.

*Proof* Let $\sum_{n=1}^{\infty} v_n$ be a rearrangement of $\sum_{n=1}^{\infty} a_n$. Then $\sum_{n=1}^{\infty} v_n$ is absolutely convergent by Theorem 3.4.2.

The positive and negative terms are separated out in the following way. Define $b_n, c_n, x_n, y_n$ by the relations

$$b_n = \begin{cases} a_n & \text{if } a_n \geqslant 0, \\ 0 & \text{if } a_n < 0, \end{cases} \qquad c_n = \begin{cases} 0 & \text{if } a_n \geqslant 0, \\ -a_n & \text{if } a_n < 0, \end{cases}$$

$$x_n = \begin{cases} v_n & \text{if } v_n \geqslant 0, \\ 0 & \text{if } v_n < 0, \end{cases} \qquad y_n = \begin{cases} 0 & \text{if } v_n \geqslant 0, \\ -v_n & \text{if } v_n < 0. \end{cases}$$

Then $\sum_{n=1}^{\infty} b_n$ contains the non-negative terms of $\sum_{n=1}^{\infty} a_n$, and $\sum_{n=1}^{\infty} x_n$ contains the non-negative terms of $\sum_{n=1}^{\infty} v_n$. Since $\sum_{n=1}^{\infty} v_n$ is a rearrangement of $\sum_{n=1}^{\infty} a_n$, it follows that $\sum_{n=1}^{\infty} x_n$ is a rearrangement of $\sum_{n=1}^{\infty} b_n$. Now $b_n \geqslant 0$ for all $n$, and $\sum_{n=1}^{\infty} b_n$ is absolutely convergent by the comparison test, since $|b_n| \leqslant |a_n|$ for all $n$ and $\sum_{n=1}^{\infty} |a_n|$ converges. By the previous theorem (Theorem 3.4.3), we see that $\sum_{n=1}^{\infty} x_n$ is absolutely convergent and

$$\sum_{n=1}^{\infty} x_n = \sum_{n=1}^{\infty} b_n. \tag{1}$$

Similarly, $\sum_{n=1}^{\infty} y_n$ is a rearrangement of $\sum_{n=1}^{\infty} c_n$. Using the same argument as before we see that $\sum_{n=1}^{\infty} c_n$ and $\sum_{n=1}^{\infty} y_n$ are both absolutely convergent and

$$\sum_{n=1}^{\infty} y_n = \sum_{n=1}^{\infty} c_n. \tag{2}$$

By definition, $a_n = b_n - c_n$ and $v_n = x_n - y_n$. From equations (1) and (2),

$$\sum_{n=1}^{\infty} v_n = \sum_{n=1}^{\infty} (x_n - y_n) = \sum_{n=1}^{\infty} x_n - \sum_{n=1}^{\infty} y_n = \sum_{n=1}^{\infty} b_n - \sum_{n=1}^{\infty} c_n$$

$$= \sum_{n=1}^{\infty} a_n.$$

The rearranged series $\sum_{n=1}^{\infty} v_n$ therefore has the same sum as the original series. Thus all rearrangements of an absolutely convergent series have the same sum as the original series.

Finally, we consider what happens if we multiply two absolutely convergent series. Using the normal rules of algebra to take out the brackets, the product

$$(a_1 + a_2 + a_3 + \ldots)(b_1 + b_2 + b_3 + \ldots)$$

of the two series $\sum_{n=1}^{\infty} a_n$ and $\sum_{n=1}^{\infty} b_n$, is the sum of terms of the form $a_r b_s$, where the sum is taken over all possible values of $r$ and $s$. These terms can be displayed in the following infinite array:

$$\begin{array}{ccccc} a_1b_1 & a_1b_2 & a_1b_3 & a_1b_4 & \ldots \\ a_2b_1 & a_2b_2 & a_2b_3 & a_2b_4 & \ldots \\ a_3b_1 & a_3b_2 & a_3b_3 & a_3b_4 & \ldots \\ a_4b_1 & a_4b_2 & a_4b_3 & a_4b_4 & \ldots \\ \ldots & \ldots & \ldots & \ldots & \ldots \\ \ldots & \ldots & \ldots & \ldots & \ldots \end{array}$$

They can be arranged in many different ways to give a series. For example, we could start in the top left-hand corner and travel along diagonals to give the series.

$$a_1b_1 + (a_1b_2 + a_2b_1) + (a_1b_3 + a_2b_2 + a_3b_1) + \ldots. \qquad (1)$$

In this case the first bracket contains terms $a_r b_s$ for which $r + s = 3$, the second terms for which $r + s = 4$, etc. In fact, if we write

$$c_n = a_1b_{n-1} + a_2b_{n-2} + \ldots + a_{n-1}b_1$$
$$= \sum_{r=1}^{n-1} a_r b_{n-r} \qquad (n \geqslant 2), \qquad (2)$$

then $c_n$ is the sum of all terms $a_r b_s$ for which $r + s = n$. The series (1) could then be written as $\sum_{n=2}^{\infty} c_n$, a form which is commonly called the Cauchy form of the product.

Another way of ordering the terms $a_r b_s$ is to start again at the top left-hand corner and then to travel along sides of a square. This would give the series

$$a_1b_1 + (a_1b_2 + a_2b_2 + a_2b_1) + (a_1b_3 + a_2b_3 + a_3b_3 + a_3b_2 + a_3b_1) + \ldots,$$

which is a very convenient order in some circumstances. For if we take the first term and the first bracket, these four terms are $(a_1 + a_2)(b_1 + b_2)$, while the first term and first two brackets contain the terms which appear in the product $(a_1 + a_2 + a_3)(b_1 + b_2 + b_3)$, and so on. The reader should remember these two special examples as these two ways of rearranging the terms $a_r b_s$ in a series will be used in the following theorem.

Initially, we arranged the terms $a_r b_s$ to produce a series $\sum_{n=2}^{\infty} c_n$, where

$$c_n = a_1b_{n-1} + a_2b_{n-2} + \ldots + a_{n-1}b_1.$$

The terms $a_r b_s$ are those which appear when we multiply together the two series $\sum_{n=1}^{\infty} a_n$ and $\sum_{n=1}^{\infty} b_n$. Is the value of $\sum_{n=2}^{\infty} c_n$ equal to the value of product $(\sum_{n=1}^{\infty} a_n)(\sum_{n=1}^{\infty} b_n)$? More generally, suppose we arrange the terms

$a_r b_s$ in some order to produce a single series, which we will denote by $\sum_{r,s=1}^{\infty} a_r b_s$. Is $\sum_{r,s=1}^{\infty} a_r b_s$ convergent? If it is convergent, is its sum equal to the value of the product $(\sum_{n=1}^{\infty} a_n)(\sum_{n=1}^{\infty} b_n)$? The next theorem supplies an answer.

THEOREM 3.4.5 Let $\sum_{n=1}^{\infty} a_n$ and $\sum_{n=1}^{\infty} b_n$ be two absolutely convergent series of real numbers. Then the series $\sum_{r,s=1}^{\infty} a_r b_s$ is absolutely convergent and

$$\sum_{r,s=1}^{\infty} a_r b_s = \left(\sum_{n=1}^{\infty} a_n\right)\left(\sum_{n=1}^{\infty} b_n\right).$$

*Proof* Let $t_n$ be the sum of the moduli of the first $n$ terms of the series $\sum_{r,s=1}^{\infty} a_r b_s$. Then there is some corresponding positive integer $M$ such that all these moduli (and probably many more besides) are contained in the square array below:

$$\begin{array}{ccccc} |a_1b_1| & |a_1b_2| & |a_1b_3| & \dots & |a_1b_M| \\ |a_2b_1| & |a_2b_2| & |a_2b_3| & \dots & |a_2b_M| \\ \dots & \dots & \dots & & \dots \\ |a_Mb_1| & |a_Mb_2| & |a_Mb_3| & \dots & |a_Mb_M|. \end{array}$$

The sum of all the terms in this square array is $(\sum_{r=1}^{M} |a_r|)(\sum_{s=1}^{M} |b_s|)$. Hence

$$t_n \leqslant \left(\sum_{r=1}^{M} |a_r|\right)\left(\sum_{s=1}^{M} |b_s|\right) = R_M S_M, \tag{1}$$

where

$$R_M = |a_1| + |a_2| + \dots + |a_M|$$
$$S_M = |b_1| + |b_2| + \dots + |b_M|.$$

Now the sequences $(R_M)$ and $(S_M)$ are both increasing. Moreover, $R_M$ and $S_M$ both tend to finite limits as $M \to \infty$, because $\sum_{n=1}^{\infty} |a_n|$ and $\sum_{n=1}^{\infty} |b_n|$ both converge. Let $R = \lim_{M\to\infty} R_M$ and $S = \lim_{M\to\infty} S_M$. Then

$$R_M \leqslant R, \qquad S_M \leqslant S$$

for all $M$. From (1), it now follows that

$$t_n \leqslant R_M S_M \leqslant RS$$

for all $n$, i.e. the sequence $(t_n)$ is bounded above. Since $t_n$ is the sum of the moduli of the first $n$ terms, $(t_n)$ is increasing. Hence $t_n$ tends to a finite limit as $n \to \infty$ and, therefore, the series $\sum_{r,s=1}^{\infty} a_r b_s$ is absolutely convergent. This guarantees that the sum is unaltered when the order of the terms is changed. In order to prove that

$$\sum_{r,s=1}^{\infty} a_r b_s = \left(\sum_{n=1}^{\infty} a_n\right)\left(\sum_{n=1}^{\infty} b_n\right)$$

we arrange the series in the following way. Start at the top left-hand corner.

Then follow the arrows as in the diagram below:

$$
\begin{array}{ccccc}
a_1b_1 & & a_1b_2 & & a_1b_3 \\
\downarrow & & \uparrow & & \uparrow \\
a_2b_1 & \rightarrow & a_2b_2 & & a_2b_3 \\
 & & & & \uparrow \\
a_3b_1 & \rightarrow & a_3b_2 & \rightarrow & a_3b_3 \\
 & & & & \\
a_4b_1 & & \cdots & & \cdots
\end{array}
$$

This gives a series beginning

$$a_1b_1 + a_2b_1 + a_2b_2 + a_1b_2 + a_3b_1 + a_3b_2 + \ldots .$$

The first $n^2$ terms of this series can be arranged in the following square array:

$$
\begin{array}{ccccc}
a_1b_1 & a_1b_2 & a_1b_3 & \ldots & a_1b_n \\
a_2b_1 & a_2b_2 & a_2b_3 & \ldots & a_2b_n \\
\ldots & \ldots & \ldots & \ldots & \ldots \\
a_nb_1 & a_nb_2 & a_nb_3 & \ldots & a_nb_n
\end{array}
$$

which has sum $T_{n^2}$ given by

$$T_{n^2} = \left( \sum_{k=1}^{n} a_k \right) \left( \sum_{l=1}^{n} b_l \right).$$

As $n \to \infty$, $T_{n^2} \to AB$, where

$$A = \sum_{k=1}^{\infty} a_k, \qquad B = \sum_{l=1}^{\infty} b_l.$$

Hence the absolutely convergent series $\sum_{r,\,s=1}^{\infty} a_rb_s$ has sum $AB$. Thus

$$\sum_{r,\,s=1}^{\infty} a_rb_s = AB = \left( \sum_{n=1}^{\infty} a_n \right) \left( \sum_{n=1}^{\infty} b_n \right)$$

and the proof is complete.

Since the series $\sum_{r,\,s=1}^{\infty} a_rb_s$ is absolutely convergent with sum $AB$, we can rearrange it and the sum remains $AB$. If we collect together all the terms $a_rb_s$ for which $r + s = n$, we obtain

$$a_1b_{n-1} + a_2b_{n-2} + \ldots + a_{n-1}b_1 \qquad (n \geqslant 2).$$

Write $\qquad c_n = a_1b_{n-1} + a_2b_{n-2} + \ldots + a_{n-1}b_1 \qquad (n \geqslant 2).$

Then we need to sum over all $n \geqslant 2$, to obtain all the terms of $\sum_{r,\,s=1}^{\infty} a_rb_s$. Thus $\sum_{r,\,s=1}^{\infty} a_rb_s$ can be rearranged so as to produce $\sum_{n=2}^{\infty} c_n$. This gives the following.

COROLLARY Let $\sum_{n=1}^{\infty} a_n$ and $\sum_{n=1}^{\infty} b_n$ be two absolutely convergent series of real numbers. For $n \geqslant 2$, write

$$c_n = a_1 b_{n-1} + a_2 b_{n-2} + \ldots + a_{n-1} b_1.$$

Then $\sum_{n=2}^{\infty} c_n$ is absolutely convergent and

$$\sum_{n=2}^{\infty} c_n = \left(\sum_{n=1}^{\infty} a_n\right)\left(\sum_{n=1}^{\infty} b_n\right).$$

The series $\sum_{n=2}^{\infty} c_n$ if often called the Cauchy form of the product in honour of the celebrated French mathematician Cauchy, who made many contributions to analysis, as the reader may already have noticed!

## APPENDIX

In Chapter 3 we discovered that the series

$$1 - \tfrac{1}{2} + \tfrac{1}{3} - \tfrac{1}{4} + \tfrac{1}{5} - \tfrac{1}{6} + \tfrac{1}{7} - \tfrac{1}{8} + \ldots$$

can be rearranged to give convergent series with different sums. In fact, it was confidently asserted that rearrangements exist with any chosen sum. This section contains a brief sketch of a method of constructing such rearrangements. Only the outline of the method is given; the details are left to the reader.

We first note that the sum $s_n$ of the first $n$ positive terms is given by

$$s_n = 1 + \frac{1}{3} + \frac{1}{5} + \frac{1}{7} + \frac{1}{9} + \ldots + \frac{1}{2n-1}. \tag{1}$$

Now,

$$\begin{aligned} s_n &> \frac{1}{2} + \frac{1}{4} + \frac{1}{6} + \frac{1}{8} + \frac{1}{10} + \ldots + \frac{1}{2n} \\ &= \frac{1}{2}\left(1 + \frac{1}{2} + \frac{1}{3} + \ldots + \frac{1}{n}\right) \\ &= \tfrac{1}{2}(\log n + \gamma_n), \end{aligned}$$

and therefore $s_n \to \infty$ as $n \to \infty$.

Similarly, if $t_n$ is the sum of the first $n$ negative terms, then

$$\begin{aligned} t_n &= -\frac{1}{2} - \frac{1}{4} - \frac{1}{6} - \frac{1}{8} - \ldots - \frac{1}{2n} \qquad (2) \\ &= -\tfrac{1}{2}(\log n + \gamma_n) \end{aligned}$$

and $t_n \to -\infty$ as $n \to \infty$. Thus if we choose any non-negative number $x$ then the sum of the first $n$ positive terms exceeds $x$, provided $n$ is sufficiently large. Admittedly, the number $n$ may need to be enormous if it is to be 'sufficiently large'. For example, if $x = 2000$, then the number $n$ would need to be very large indeed to ensure that $s_n > x$; but this doesn't matter. What is really important is that such a number $n$ exists.

Similarly, if $x < 0$ then we can choose a sufficiently large number of negative terms so that the sum is less than $x$.

Suppose now that we choose any non-negative real number $x$ and we wish to develop a method for constructing a series with sum $x$. We begin the series by writing down the first $n$ positive terms, where $n$ is chosen so that the sum $s_n$ (see equation (1)) exceeds $x$, but $s_{n-1} \leqslant x$. For example, if the desired sum $x$ is $\pi/4$, then $n = 1$ and the series starts with just the first positive term 1. However, if $x = 1$ then we need the first two positive terms $1 + \frac{1}{3}$, if the sum is to *exceed* $x$.

These first $n$ positive terms are then followed by the first $m$ negative terms, where $m$ is chosen so that the total sum $s_n + t_m$ of all $m + n$ terms is less than $x$, but $s_n + t_{m-1} \geqslant x$, i.e. we include just enough negative terms so that the total sum is less than $x$. For example, if the required sum $x$ is $\pi/4$, then $m = 1$ and the series starts $1 - \frac{1}{2}$, whereas if $x = 1$ then $n = 2, m = 1$ and the series commences $1 + \frac{1}{3} - \frac{1}{2}$. However, if the desired sum $x = \frac{1}{10}$, then $m = 3$ and the series begins with the terms

$$1 - \tfrac{1}{2} - \tfrac{1}{4} - \tfrac{1}{6}.$$

Now that the sum is less than $x$, we continue moving along the series, taking positive terms and writing them down. As each one joins the new series, we check the total sum and stop as soon as the total sum again exceeds $x$. It is then time to include the next negative terms. We continue in this manner, including blocks of positive terms and blocks of negative terms. Each time we stop adding positive terms as soon as the sum exceeds $x$ and we stop including negative terms just as soon as the total sum is less than $x$. This gives a recipe for constructing a rearrangement with sum $x$. Naturally, if $x < 0$, then we need to start with negative terms.

## *Examples*

The examples below show the first few terms of the rearrangements produced by the above method for given values of $x$.

1. $x = \pi/4$. The rearranged series begins

$$1 - \tfrac{1}{2} + \tfrac{1}{3} - \tfrac{1}{4} + \tfrac{1}{5} + \tfrac{1}{7} - \tfrac{1}{6} + \tfrac{1}{9} - \tfrac{1}{8} + \tfrac{1}{11} - \tfrac{1}{10} + \ldots$$

2. $x = 0$. The rearranged series begins

$$1 - \tfrac{1}{2} - \tfrac{1}{4} - \tfrac{1}{6} - \tfrac{1}{8} + \tfrac{1}{3} - \tfrac{1}{10} - \tfrac{1}{12} - \tfrac{1}{14} - \tfrac{1}{16} + \tfrac{1}{5} - \ldots$$

3. $x = \pi/3$. The rearranged series begins

$$1 + \tfrac{1}{3} - \tfrac{1}{2} + \tfrac{1}{5} - \tfrac{1}{4} + \tfrac{1}{7} + \tfrac{1}{9} - \tfrac{1}{6} + \tfrac{1}{11} + \tfrac{1}{13} - \tfrac{1}{8} + \tfrac{1}{15} + \ldots$$

4. $x = -\frac{1}{12}$. The rearranged series begins

$$-\tfrac{1}{2}+1-\tfrac{1}{4}-\tfrac{1}{6}-\tfrac{1}{8}-\tfrac{1}{10}+\tfrac{1}{3}-\tfrac{1}{12}-\tfrac{1}{14}-\tfrac{1}{16}-\tfrac{1}{18}-\tfrac{1}{20}+\tfrac{1}{5}-\ldots$$

5. $x = 1$. The rearranged series begins

$$1+\tfrac{1}{3}-\tfrac{1}{2}+\tfrac{1}{5}-\tfrac{1}{4}+\tfrac{1}{7}+\tfrac{1}{9}-\tfrac{1}{6}+\ldots$$

---

It can be shown that this construction gives a convergent series whose sum is $x$. The following exercises provide a guide for the reader who is brave enough to wish to embark on a proof.

## *EXERCISES*

In all of the following exercises it is assumed that the above construction is being used to produce a rearrangement of the series

$$1-\tfrac{1}{2}+\tfrac{1}{3}-\tfrac{1}{4}+\tfrac{1}{5}-\tfrac{1}{6}+\tfrac{1}{7}-\tfrac{1}{8}+\ldots$$

with sum $x$.

**1** When $x \geqslant 0$, the rearranged series begins with a block of $n$ positive terms. Verify that

$$0 < s_n - x < \frac{1}{2n-1},$$

where $s_n$ is the sum of these $n$ positive terms.

**2** Check that for **all** real values of $x$ the difference between $x$ and the sum of all the terms in the first block does not exceed the value of the modulus of the last term in the block.

Check also that the difference between $x$ and the sum of all the terms in the first $N$ blocks does not exceed the value of the modulus of the last term in the $N$th block.

**3** (Harder) Let $t_k$ be the sum of the first $k$ terms of the rearranged series. Suppose the number of terms in the first $N$ blocks is less than or equal to $k$ (i.e. all the terms of the first $N$ blocks, and possibly many more besides, are included in the first $k$ terms). Show that

$$|t_k - x| < \frac{1}{N-1} \qquad (N \geqq 2).$$

(Take care! If $x < 0$ then the series begins with a block of negative terms. In this case, the term 1 must be in the second block etc.)

Deduce that the rearranged series converges and has sum $x$.

## TESTS FOR CONVERGENCE—SUMMARY

Comparison test (p. 71)

Limit form of comparison test (p. 73)

D'Alembert's test (ratio test) (p. 76)

Integral test (p. 83)

Alternating series test (p. 89)

## *MISCELLANEOUS EXERCISES 3*

Decide whether the following series converge or diverge:

**1** $\sum_{n=1}^{\infty} \frac{(3n)!3^{-n}}{n!(2n)!}$; **2** $\sum_{n=1}^{\infty} \left(\frac{-n}{n+1}\right)^n$; **3** $\sum_{n=1}^{\infty} \frac{\sin n}{n^2}$;

**4** $\sum_{n=1}^{\infty} \frac{1}{\sqrt{n}} \tan\left(\frac{(2n-1)\pi}{4}\right)$; **5** $\sum_{n=1}^{\infty} \frac{n^2+2}{3n^3+4n}$; **6** $\sum_{n=2}^{\infty} \frac{1}{n \log n}$;

**7** $\sum_{n=1}^{\infty} (-1)^n n^{1/n}$; **8** $\sum_{n-1}^{\infty} (-1)^n \sin\left(\frac{1}{\sqrt{n}}\right)$; **9** $\sum_{n=1}^{\infty} \frac{(5n)!}{(30)^n(3n)!(2n)!}$;

**10** $\sum_{n=1}^{\infty} \frac{(n^2+1)^4}{(n^3+1)^3}$; **11** $\sum_{n=1}^{\infty} \frac{2+(-1)^n}{n^{3/2}}$; **12** $\sum_{n=1}^{\infty} \frac{n^2+1}{4n^4+n^2-n-1}$;

**13** $\sum_{n=1}^{\infty} \frac{(2n)!}{(n!)^2}$; **14** $\sum_{n=1}^{\infty} \frac{\sin(n!)}{n^2+n+1}$; **15** $\sum_{n=1}^{\infty} \frac{(-1)^n n^2}{n^3+1}$;

**16** $\sum_{n=1}^{\infty} \frac{9^n(3n)!n!}{(4n)!}$; **17** $\sum_{n=1}^{\infty} \frac{2+(-1)^n}{\sqrt{n}}$; **18** $\sum_{n=1}^{\infty} \sin\left(\frac{\pi n}{2}\right)\cos\left(\frac{\pi}{n}\right)$;

**19** $\sum_{n=1}^{\infty} \frac{n^2+n+1}{\sqrt{n^7+n^4+1}}$; **20** $\sum_{n=1}^{\infty} \sin\left(\frac{(-1)^n}{n}\right)$; **21** $\sum_{n=1}^{\infty} \frac{1}{n(\log n)^{3/2}}$.

## HINTS FOR SOLUTION OF EXERCISES

### Exercises 3.2.1

**2** Use $\log\left(1+\frac{1}{n}\right) = \log[(n+1)/n] = \log(n+1) - \log n$.

# ANSWERS TO EXERCISES

## Exercises 3.2.1

**1** $\frac{1}{4}$. 2 $s_n = \log(n+1)$.

## Exercises 3.3.1

**3** In each case the comparison test or the limit form of the comparison test is used. The series $\sum_{n=1}^{\infty} b_n$ used for comparison is given in brackets.

(a) Diverges ($b_n = 1/n$). (b) Converges ($b_n = 1/n^2$).
(c) Converges ($b_n = 1/n^2$). (d) Diverges ($b_n = 1/n$).
(e) Converges (use $1/(n^{n+1/n}) < 1/n^2$ for all $n \geqslant 2$).
(f) Diverges ($b_n = 1/n^{1/3}$).

## Exercises 3.3.2

**3** In parts (a) and (f) d'Alembert's test is used. The comparison test or its limit form are used for all the rest. The series $\sum_{n=1}^{\infty} b_n$ used for comparison is given brackets.

(a) Converges (d'Alembert's test or comparison test with $b_n = (\frac{5}{7})^n$).
(b) Converges ($b_n = 1/n^2$). (c) Diverges ($b_n = n^5$).
(d) Diverges ($b_n = 1/\sqrt{n}$).
(e) Diverges ($b_n = 1/n$). Remember that $(1 + 1/n)^n \to \mathrm{e}$ as $n \to \infty$.
(f) Converges.
(g) Converges ($b_n = 1/n^2$). Remember that $|\sin n| \leqslant 1$ for all $n$.

## Exercises 3.3.3

**3** (a) Converges (alternating series test, $\sin[(n+1)\pi/n] = \sin(n\pi + \pi/n) = (-1)^n \sin(\pi/n)$).
(b) Converges (comparison test, $b_n = 1/n\sqrt{n}$).
(c) Diverges ($a_n \nrightarrow 0$ as $n \to \infty$). (d) Converges (integral test).
(e) Diverges ($a_n \nrightarrow 0$ as $n \to \infty$). (f) Converges (d'Alembert's test).

## Miscellaneous Exercises 3

**1** Diverges (d'Alembert's test).

**2** Diverges ($a_n \nrightarrow 0$ as $n \to \infty$).

**3** Converges (comparison test with $b_n = 1/n^2$ proves series is absolutely convergent).

**4** Converges (alternating series test and $\tan[(2n-1)\pi/4] = (-1)^{n+1}$).

**5** Diverges (comparison test, $b = 1/n$).

**6** Diverges (integral test).

**7** Diverges ($a_n \nrightarrow 0$ as $n \to \infty$).

**8** Converges (alternating series test).

**9** Converges (d'Alembert's test).

**10** Diverges (comparison test, $b_n = 1/n$).

**11** Converges (comparison test, $b_n = 1/n\sqrt{n}$; use the fact that $(2+(-1)^n)/n^{3/2} \leqslant 3/n^{3/2}$).

**12** Converges (comparison test, $b_n = 1/n^2$).

**13** Diverges (d'Alembert's test).

**14** Converges (comparison test with $b_n = 1/n^2$ proves that the series is absolutely convergent).

**15** Converges (alternating series test).

**16** Converges (d'Alembert's test).

**17** Diverges (comparison test, $b_n = 1/\sqrt{n}$. Use $[2+(-1)]/\sqrt{n} \geqslant 1/\sqrt{n}$).

**18** Diverges ($a_{2n+1} = (-1)^{n+1}\cos(\pi/n)$ and $a_n \nrightarrow 0$ as $n \to \infty$).

**19** Converges (comparison test, $b_n = 1/n\sqrt{n}$).

**20** Converges (alternating series test; use the fact that $\sin[(-1)^n/n] = (-1)^n \sin(1/n)$).

**21** Converges (integral test).

# 4 FUNCTIONS
## Limits of functions, continuous functions

### 4.1 LIMITS OF FUNCTIONS

The principal characters in this chapter are real-valued functions. For the benefit of any reader whose knowledge of functions may be a little shaky, we recall some basic ideas. As we are only concerned with certain types of functions, the definitions are restricted to those of interest to us, viz. the real valued functions. (*Note:* A more comprehensive treatment of functions is contained in the companion volume in this series, *Guide to Abstract Algebra.*) First suppose $X$ and $Y$ are both sets of real numbers. Suppose also that there is some rule which assigns to each real number $x \in X$ a unique real number $y \in Y$. Then this rule defines a function from $X$ into $Y$. If we use the letter $f$ for the function then we would write $f: X \to Y$ to signify that the function is from $X$ into $Y$. Sometimes it is convenient to think in terms of the diagram shown as Fig. 4.1.

If the number $y \in Y$ corresponds to the number $x \in X$, then we write $y = f(x)$. The set $X$ is called the **domain of definition** of $f$ (sometimes abbreviated to the 'domain of $f$') and the phrase '$f(x)$ is defined' means $x \in X$, i.e. $x$ is in the domain of definition of $f$ and so $f(x)$ has a definite value. The set of all numbers $y \in Y$ such that $y = f(x)$ for some $x \in X$ is denoted by $f(X)$ and is called the **range** of $f$. Thus the range of $f$ is the set $f(X)$, where

$$f(X) = \{y \in Y : y = f(x) \text{ for some } x \in X\}.$$

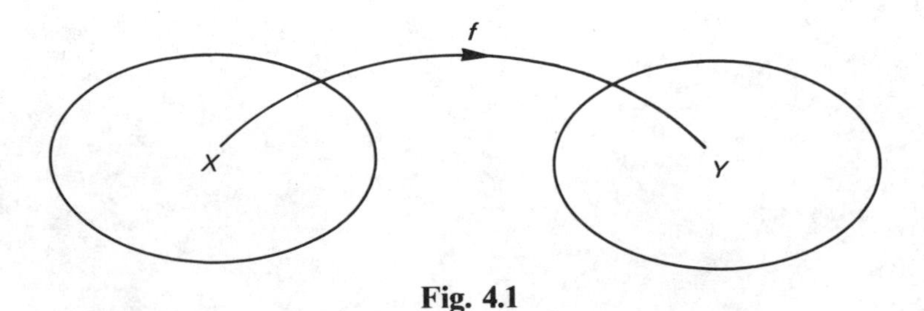


**Fig. 4.1**

For any $x \in X$, the number $f(x)$ is called the **image** of $x$. The set $f(X)$ is therefore the set of images of all numbers in $X$. This set $f(X)$, of course, need not be the whole of the set $Y$. All the definition stipulates is that every element of the form $f(x)$ $(x \in X)$ belongs to $Y$; it does not require that these are the only members of $Y$. In practice, therefore, it is often convenient to use the whole of $\mathbb{R}$ as the set $Y$. The majority of functions we consider will be treated as functions $f: X \to \mathbb{R}$. Occasionally, in particular circumstances, we will need to use a set $Y$ which does not include all the real numbers, but such instances will be rare in this volume. Below we give a few examples of functions.

*Examples 4.1.1*

1. The function $f: \mathbb{R} \to \mathbb{R}$ is given by

$$f(x) = x^2$$

for all $x \in \mathbb{R}$.

2. The function $f: \mathbb{R} \to \mathbb{R}$ is given by

$$f(x) = \begin{cases} 0 & \text{if } x \text{ is rational,} \\ 1 & \text{if } x \text{ is irrational.} \end{cases}$$

3. Let $X$ be the set of all positive real numbers. The function $f: X \to \mathbb{R}$ is given by

$$f(x) = \begin{cases} 0 & \text{if } x \text{ is irrational and } x > 0, \\ 1/q & \text{if } x \text{ is rational and } x = p/q, \\ & \text{where } p, q \text{ are positive integers with no} \\ & \text{common factor greater than 1.} \end{cases}$$

(The condition that $p, q$ have no common factors means that the fraction $p/q$ has been cancelled down to its lowest terms.)

4. The function $f: \mathbb{R}\backslash\{0\} \to \mathbb{R}$ is defined by

$$f(x) = 1/x \qquad (x \in \mathbb{R},\ x \neq 0).$$

In Example 1, we notice that $x^2 \geqslant 0$ for all real numbers $x$ and so the range is the set of all non-negative real numbers. In Example 2, $f(x)$ is either 0 or 1 and the range is the set $\{0, 1\}$ with just two elements. The function in Example 3, is at first sight a little strange. It satisfies the rules for a function since there is precisely one corresponding value $f(x)$ assigned to each positive real number $x$ and so it is undoubtedly a function. Its range consists of all the rational numbers $1/q$ for which $q$ is a positive integer, together with the number 0. Example 4 is left to the reader.

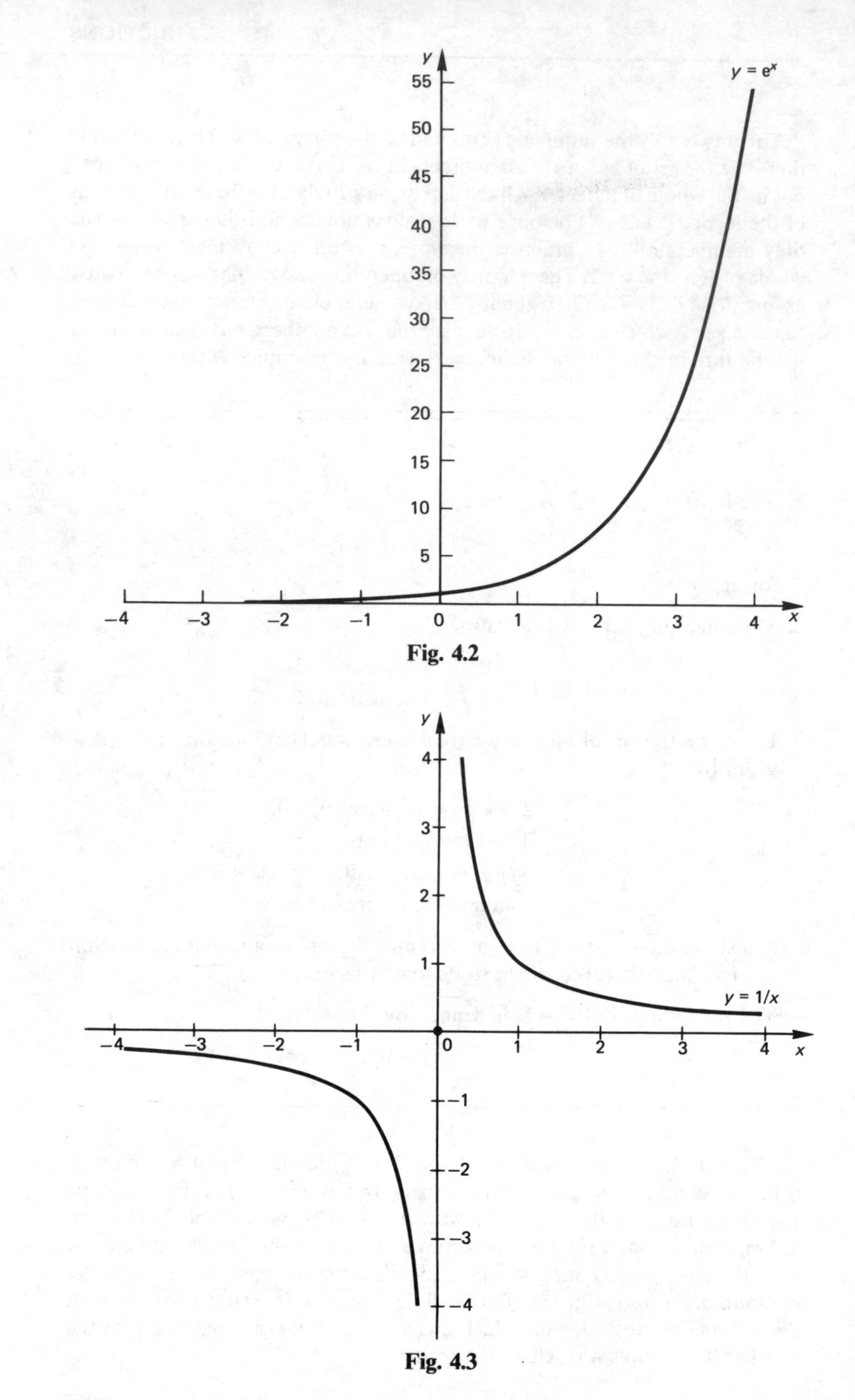
y
55
50
45
40
35
30
25
20
15
10
5
y = e^x
−4
−3
−2
−1
0
1
2
3
4
x

**Fig. 4.2**

y
4
3
2
1
−1
−2
−3
−4
y = 1/x
−4
−3
−2
−1
0
1
2
3
4
x

**Fig. 4.3**

It will not have escaped the reader's attention that Examples 1 and 4 are familiar functions whose graphs are easy to sketch, whereas Example 3 illustrates that there are functions whose graphs cannot be sketched (even with a considerable amount of ingenuity). We must bear this in mind in the rest of the chapter. Our concepts must be framed in such a way that they can be used with functions whose graphs cannot be drawn. They must, therefore, be couched in terms of symbols rather than pictures, although we will use sketches to give us the initial ideas.

Throughout the remainder of the chapter we will be considering functions $f: X \to \mathbb{R}$, whose domain of definition $X$ contains an interval (or intervals). It will probably simplify the discussion if we now introduce the required notation for intervals. In evolving the notation we use two letters $a, b$ to denote arbitrary real numbers.

**Intervals: notation**

1. The set of all real numbers $x$ such that $x > a$ is denoted by $(a, \infty)$.
2. The set of all real numbers $x$ such that $x \geqslant a$ is denoted by $[a, \infty)$.
3. The set of all real numbers $x$ such that $x < b$ is denoted by $(-\infty, b)$.
4. The set of all real numbers $x$ such that $x \leqslant b$ is denoted by $(-\infty, b]$.
5. Let $a, b$ be real numbers with $a < b$. The set of all real numbers $x$ such that $a < x < b$ is denoted by $(a, b)$ and is called an *open interval.*
6. Let $a, b$ be real numbers with $a < b$. The set of all real numbers $x$ such that $a \leqslant x \leqslant b$ is denoted by $[a, b]$ and it is called *a closed interval.*

We now turn to the question of limits of a function.

To give us some ideas on how to tackle the problem we begin by sketching the graphs of some fairly elementary functions. These are shown as Figs 4.2 to 4.6.

It seems intuitively obvious that (a) $e^x \to \infty$ as $x \to \infty$, (b) $1/x \to 0$ as $x \to \infty$, (c) $1/(1 + x^2) \to 0$ as $x \to \infty$ and (d) $(x^2 - 1)/(x^2 + 1) \to 1$ as $x \to \infty$. The last example may cause a little hesitation. The graph seems to suggest that $x + \sin x \to \infty$ as $x \to \infty$. A moments thought suffices to confirm this conjecture, since $x + \sin x \geqslant x - 1$ for all $x$ and quite obviously $x - 1 \to \infty$ as $x \to \infty$. But is not sufficient just to sketch graphs. The functions we have used as illustrations had graphs which could be drawn without undue difficulty. There are, however, functions whose graphs are very difficult to draw and functions whose graphs are impossible to draw. Our definitions of concepts like the limit of a function must be applicable irrespective of whether the graph of the function can be sketched or not. The definitions we eventually give must therefore be in terms of symbols. The form in which they are cast has not changed over the last century. In fact, they are essentially the same as those given by Heine in 1872.

Let us now return to our examples. It seemed obvious that $e^x \to \infty$ as $x \to \infty$ and $x + \sin x \to \infty$ as $x \to \infty$. What properties of the functions led to this conclusion? Clearly $e^x$ increases as $x$ increases and there is no limit on

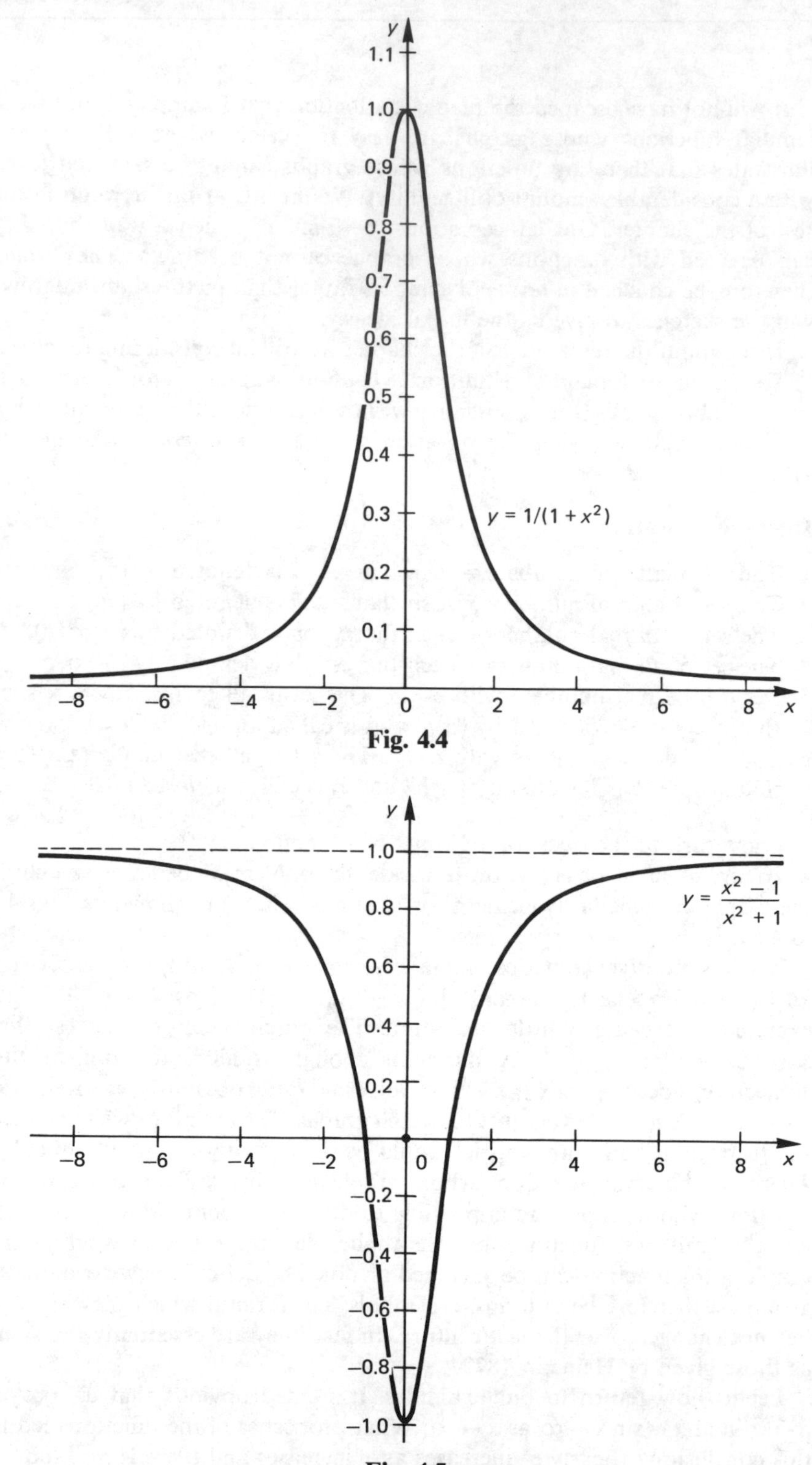
y
1.1
1.0
0.9
0.8
0.7
0.6
0.5
0.4
0.3
0.2
0.1
y = 1/(1 + x^2)
−8
−6
−4
−2
0
2
4
6
8
x
Fig. 4.4

y
1.0
0.8
0.6
0.4
0.2
−0.2
−0.4
−0.6
−0.8
−1.0
y = (x^2 − 1)/(x^2 + 1)
−8
−6
−4
−2
0
2
4
6
8
x
Fig. 4.5

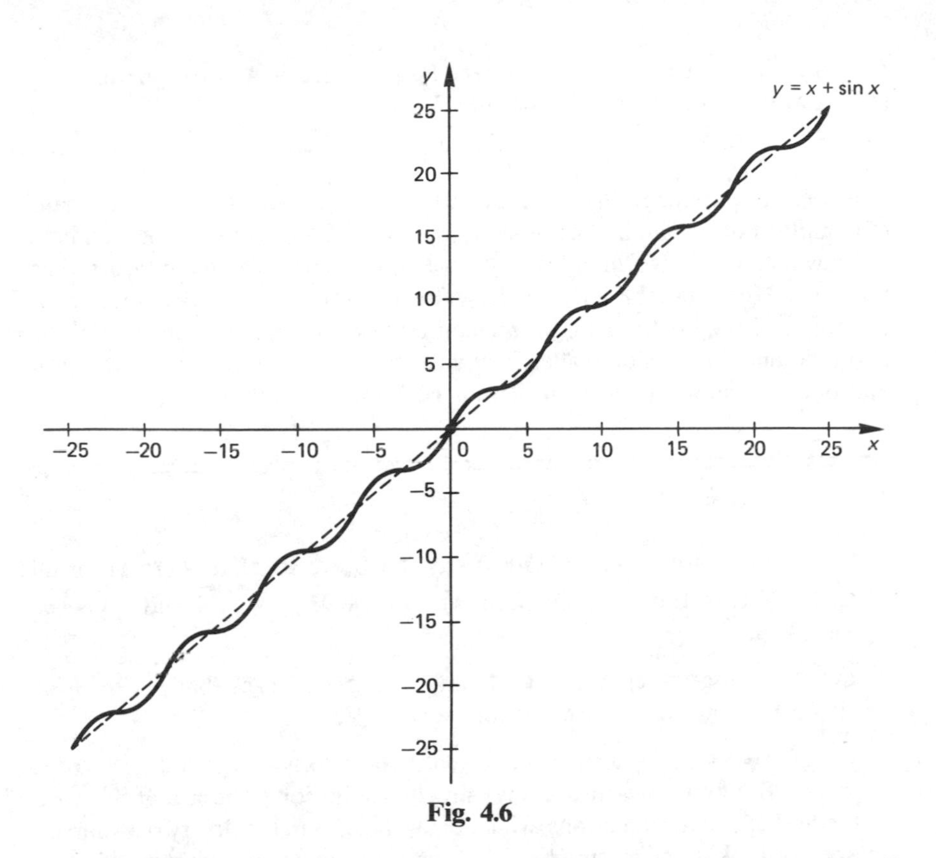


**Fig. 4.6**

its size i.e. there is no upper bound on the values assumed by $e^x$. In fact, if we choose any positive real number $A$ (however large) then $e^x > A$ for all sufficiently large $x$, i.e. $e^x$ remains greater than $A$ for all sufficiently large $x$. Since $e^x$ is a fairly easy function to handle we can be a little more precise. Let $A$ be any positive real number. Then $e^x > A$ whenever $x > \log A$. If we write $X_1 = \log A$, then we have $e^x > A$ for all $x > X_1$. Could we make a similar statement about the last example? The answer is yes. For $x + \sin x > A$ whenever $x > A + 1$. Suppose that $X_1 = A + 1$. Then if $x > X_1$ we have

$$x + \sin x \geqslant x - 1 > X_1 - 1 = A.$$

Since the graph fluctuates a little locally, there will also be values of $x$ with $x \leqslant A + 1$ for which $x + \sin x > A$. But this doesn't matter. The vital property is that the graph remains above the line $y = A$ for all sufficiently large values of $x$ (in this case $x > A + 1$ will do). This is just the property we need to develop our definition. We will, of course, need to assume that the function $f$ has the property that $f(x)$ has a value for all very large values of $x$.

DEFINITION 4.1.1 Let $f$ be a real valued function such that $f(x)$ is defined for all $x > a$ (where $a$ is some given real number). Then we say that $f(x) \to \infty$

as $x \to \infty$ if given any $A > 0$ (however large), there is a corresponding $X_1$ (whose value depends on $A$) such that $X_1 > a$ and

$$f(x) > A \qquad \text{for } \textbf{all } x > X_1.$$

*Note:* The requirement that $f(x)$ is defined for all $x > a$ means that the domain of definition of $f$ contains the interval $(a, \infty)$, i.e. it contains all the numbers $x$ for which $x > a$. Normally the symbol $X$ is used in the definition rather than $X_1$. However, the letter $X$ has already appeared for the domain of definition of the function, and it seemed unwise to use the same symbol for two different things. The reader may wish to revert to the normal notation and omit the subscript 1 when there is no danger of confusion.

---

*Examples 4.1.2*

1. Let $f(x) = x^2$ for all real $x$. Given any $A > 0$, we see that $f(x) > A$ for all $x > +\sqrt{A}$. In this case, therefore, we can use $X_1 = +\sqrt{A}$ and $f(x) \to \infty$ as $x \to \infty$.

2. Let $f(x) = \log(\log x)$ for $x > 1$. Given any $A > 0$, we see that $f(x) > A$ for all $x > e^{e^A}$ and in this case we can use $X_1 = e^{e^A}$.

3. Let $f(x) = x^5 + x^4 + x \sin x + 1$. Then it looks obvious that $f(x) \to \infty$ as $x \to \infty$. But how do we prove this using the definition? The straightforward method of solving equations, which proved useful in the first two examples, doesn't look as promising here. In this case the problem is simplified by using some inequalities.

   We first notice that if $x > 1$ then

$$x^3 + \sin x \geqslant x^3 - 1 > 0,$$

   and therefore

$$x^4 + x \sin x = x(x^3 + \sin x) > 0.$$

   Thus, if $x > 1$ then

$$f(x) = x^5 + x^4 + x \sin x + 1 > x^5 + 1 > x^5.$$

   Given any $A > 0$; $x^5 > A$ for all $x > A^{1/5}$. Let $X_1$ be the larger of the two numbers 1, $A^{1/5}$, i.e. $X_1 = \max\{1, A^{1/5}\}$. Then for all $x > X_1$,

$$f(x) > x^5 > A.$$

   By definition $f(x) \to \infty$ as $x \to \infty$.

---

We notice that the working was simplified, in the above example, by the use of inequalities. The reason is not hard to understand. The graph of $y = f(x)$

is above the graph of $y = x^5$ for all $x > 1$, since $f(x) > x^5$ for all $x > 1$. We know what happens to the lower graph, since $x^5 \to \infty$ as $x \to \infty$. It follows that $f(x) \to \infty$ as $x \to \infty$.

Naturally, this idea can be generalised. Suppose we have two functions $f, g$ with the property that the graph of $f$ lies above the graph of $g$, i.e. $f(x) \geqslant g(x)$. If $g(x) \to \infty$ as $x \to \infty$, then it is clear that $f(x) \to \infty$ as $x \to \infty$. Of course, it is not really necessary to be able to draw the graphs. The inequality $f(x) \geqslant g(x)$, together with the assumption that $g(x) \to \infty$ as $x \to \infty$, is sufficient to guarantee that $f(x) \to \infty$ as $x \to \infty$. A result like this is of such practical value that it merits incorporation in a theorem.

THEOREM 4.1.1 Let $f, g$ be real valued functions such that $f(x)$ and $g(x)$ are defined for all $x > a$ and satisfy the inequality

$$f(x) \geqslant g(x) \qquad (x > a),$$

where $a$ is some given real number. If $g(x) \to \infty$ as $x \to \infty$, then $f(x) \to \infty$ as $x \to \infty$.

*Proof* Let $A$ be any positive real number. Since $g(x) \to \infty$ as $x \to \infty$, there is some $X_1$ (with $X_1 > a$) such that

$$g(x) > A \qquad \text{for all } x > X_1.$$

Thus for $x > X_1$

$$f(x) \geqslant g(x) > A,$$

and, therefore $f(x) \to \infty$ as $x \to \infty$.

---

*Example 4.1.3*

Let

$$f(x) = x^2 + x + \sin x,$$

$$g(x) = x^2.$$

For $x > 1$, $\quad f(x) = x^2 + x + \sin x \geqslant x^2 + x - 1 > x^2 = g(x)$

and therefore $f(x) \to \infty$ as $x \to \infty$ because $g(x) \to \infty$ as $x \to \infty$.

---

The reader may notice that there is a great similarity between the ideas in section 2.2 on sequences and this section on limits of functions. For example there is a great resemblance between Definitions 2.2.2 and 4.1.1, and between Theorems 2.2.1 and 4.1.1. This is not accidental. The basic ideas are to a great extent the same and the reader may find it helpful to refer back to Chapter 2 when studying this section.

## *EXERCISES 4.1.1*

**1** Use Definition 4.1.1 to show that $x^3 \to \infty$ as $x \to \infty$ (i.e. show that for each $A > 0$ there is a corresponding $X_1$ with the required properties).

**2** Use Definition 4.1.1 to show that, $x^\alpha \to \infty$ as $x \to \infty$ for each real number $\alpha > 0$.

**3** Show that $f(x) \to \infty$ as $x \to \infty$ for each of the following functions $f$:

(a) $f(x) = x^3 + x^2 + x\cos x$;
(b) $f(x) = x\log x \ (x > 0)$.

We know what kind of picture to expect if $f(x) \to \infty$ as $x \to \infty$. If for each $A > 0$ (however large), the graph of $y = f(x)$ remains above the line $y = A$ for all sufficiently large $x$, then $f(x) \to \infty$ as $x \to \infty$. But what type of behaviour do we need if $f(x) \to -\infty$? If $A > 0$, then the line $y = -A$ is below the axis. The necessary requirement is that the graph of $y = f(x)$ should remain below the line $y = -A$ for all sufficiently large $x$, i.e. $f(x) < -A$ for all sufficiently large $x$. Now the formulation of a precise definition in symbols causes no problems.

DEFINITION 4.1.2 Suppose $f(x)$ is defined for all $x > a$, where $a$ is some given real number. Then $f(x) \to -\infty$ as $x \to \infty$, if given any $A > 0$ (however large) there exists a corresponding $X_1$ (with $X_1 > a$) such that

$$f(x) < -A \qquad \text{for all } x > X_1.$$

A quick glance at Definitions 4.1.1 and 4.1.2 will immediately convince the reader that $f(x) \to -\infty$ as $x \to \infty$ if and only if $-f(x) \to \infty$ as $x \to \infty$.

Obviously Theorem 4.1.1 has an analogue for functions with limit $-\infty$. In this case we want to use a function $g$ whose graph is above that of $f$ and such that $g(x) \to -\infty$ as $x \to \infty$. This gives the following result.

THEOREM 4.1.2 Let $f, g$ be real valued functions such that $f(x)$ and $g(x)$ are defined for all $x > a$ and

$$f(x) \leqslant g(x)$$

for all $x > a$, where $a$ is some given real number. If $g(x) \to -\infty$ as $x \to \infty$, then $f(x) \to -\infty$ as $x \to \infty$.

The proof is left as an exercise for the reader.

## *EXERCISE 4.1.2*

Write out a proof for Theorem 4.1.2 (*Hint:* Either start with Definition 4.1.2 and construct a proof on similar lines to that of Theorem 4.1.1 or apply the conclusions of Theorem 4.1.1 to the functions $-f$ and $-g$ and use the fact that $f(x)\to -\infty$ as $x\to\infty$ if and only if $-f(x)\to\infty$ as $x\to\infty$.)

We recall that some of the examples near the beginning of the chapter had infinite limits whilst others had finite limits. For example, it seemed intuitively obvious that $\frac{1}{x}\to 0$ as $x\to\infty$, and $(x^2-1)/(x^2+1)\to 1$ as $x\to\infty$. How can we set about writing down a precise definition in terms of symbols for finite limits? Perhaps a glance at Chapter 2 will give us some ideas. Let us begin by looking at Fig. 2.3. This shows a sequence $(a_n)$ with the property that $a_n\to 1$ as $n\to\infty$. In this case every band about $y=1$ (however narrow) contains all the elements $a_n$ for which $n$ is sufficiently large. Normally, the band width is denoted by $2\varepsilon$ and so we have all the elements $a_n$ for which $n$ is sufficiently large between $1-\varepsilon$ and $1+\varepsilon$. In general, if the limit is $l$ then all the elements $a_n$ for which $n$ is sufficiently large lie between $l-\varepsilon$ and $l+\varepsilon$ (see Definition 2.2.4). The same ideas carry over to functions and are used to define the meaning of $f(x)\to l$ as $x\to\infty$. We now require that for all $\varepsilon>0$ (however small) the graph of $y=f(x)$ lies between the lines $y=l-\varepsilon$ and $y=l+\varepsilon$ for all sufficiently large $x$, i.e.

$$l-\varepsilon<f(x)<l+\varepsilon \tag{1}$$

for all sufficiently large $x$. For example, if $f(x)=\frac{1}{x}$ $(x>0)$, then, given $\varepsilon>0$,

$$-\varepsilon<f(x)<\varepsilon$$

for all $x>\frac{1}{\varepsilon}$. This is of the same form as (1) with $l=0$, which is not really suprising since it appears that $\frac{1}{x}\to 0$ as $x\to\infty$ (see Fig. 4.7). It is fairly clear that however narrow this band is chosen to be (i.e., however small we chose $\varepsilon$), the graph of $y=\frac{1}{x}$ lies inside the band for all sufficiently large values of $x$. We are now in a position to give a formal definition. In the statement of the definition the relation $l-\varepsilon<f(x)<l+\varepsilon$ will be expressed as $|f(x)-l|<\varepsilon$.

DEFINITION 4.1.3 Suppose $f(x)$ is defined for all $x>a$, where $a$ is some given real number. Then $f(x)\to l$ as $x\to\infty$ if given any $\varepsilon>0$ (however small) there is a corresponding $X_1$ (with $X_1>a$) such that

$$|f(x)-l|<\varepsilon$$

for **all** $x>X_1$.

In each of the Definitions 4.1.1 to 4.1.3 the concept of a limit is defined subject to the condition that $f(x)$ is defined for all $x>a$ (where $a$ is some given real number). Throughout this volume phrases like $f(x)\to l$ will be used only when such a condition is satisfied. Thus, if it is stated that $f(x)\to l$ as $x\to\infty$, then the reader knows that there is some real number $a$ such that $f(x)$ is defined for all $x>a$.

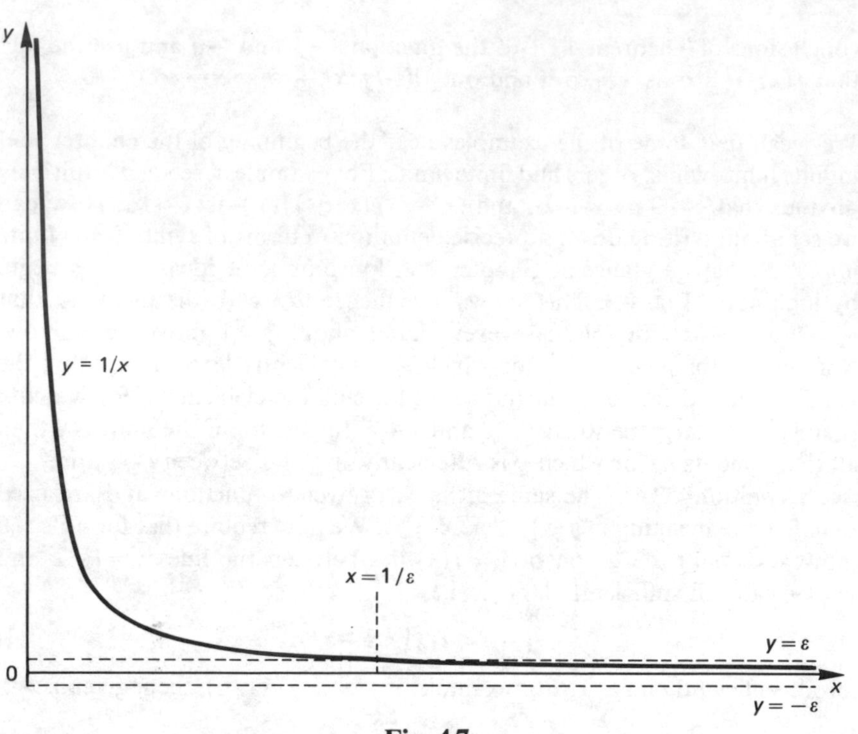


**Fig. 4.7**

*Examples 4.1.4*

1. We have seen that given any $\varepsilon > 0$,

$$-\varepsilon < \frac{1}{x} < \varepsilon$$

for all $x > \frac{1}{\varepsilon}$,

i.e.

$$\left|\frac{1}{x}\right| < \varepsilon$$

for all $x > \frac{1}{\varepsilon}$. Hence $\frac{1}{x} \to 0$ as $x \to \infty$.

2. Given any $\varepsilon > 0$, we see that

$$\left|\frac{\sin x}{x}\right| \leqslant \frac{1}{|x|} < \varepsilon$$

for all $x > \frac{1}{\varepsilon}$. Hence $(\sin x)/x \to 0$ as $x \to \infty$. In this example and the previous one $X_1 = \frac{1}{\varepsilon}$. As we already know, the value of $X_1$ depends on the value of $\varepsilon$, and the value of $X_1$ appears to increase as $\varepsilon$ gets smaller, which is exactly what we would expect.

3. For all $x$,

$$\frac{x^2-1}{x^2+1} = 1 - \frac{2}{x^2+1}.$$

Thus, given any $\varepsilon > 0$,

$$\left|\frac{x^2-1}{x^2+1} - 1\right| < \frac{2}{x^2+1} < \varepsilon,$$

provided $x^2+1 > \frac{2}{\varepsilon}$. Thus the inequality is certainly satisfied for all $x > 0$ such that $x^2+1 > \frac{2}{\varepsilon}$, i.e. it is satisfied for all sufficiently large $x$. Hence $(x^2-1)/(x^2+1) \to 1$ as $x \to \infty$.

---

## EXERCISES 4.1.3

**1** Using Definition 4.1.3, show that $1/(x^2+1) \to 0$ as $x \to \infty$.

**2** Show that if $f(x) \to \infty$ as $x \to \infty$, then $1/f(x) \to 0$ as $x \to \infty$. Deduce that $1/x^\alpha \to 0$ as $x \to \infty$ for all real numbers $\alpha > 0$.

Many of the theorems given in Chapter 2 for sequences have obvious analogues for functions. For example, a function cannot have more than one limit.

THEOREM 4.1.3 Let $f(x)$ be defined for all $x > a$, where $a$ is some given real number. Then $f(x)$ cannot tend to two different limits as $x \to \infty$.

*Proof* We consider several cases separately.

(a) *$f(x)$ cannot tend to a finite limit and an infinite limit as $x \to \infty$* Suppose $f(x) \to l$ as $x \to \infty$. Then, using Definition 4.1.3 (with the special case $\varepsilon = 1$), we see that there is a number $X_1$ (with $X_1 > a$) such that

$$|f(x) - l| < 1$$

for all $x > X_1$,

i.e. $$l - 1 < f(x) < l + 1$$

for all $x > X_1$. Since $f(x) < l + 1$ $(x > X_1)$, $f(x) \not\to \infty$ as $x \to \infty$. Further, $f(x) \not\to -\infty$ as $x \to \infty$ because $f(x) > l - 1$ for all $x > X_1$. Thus $f(x)$ cannot have a limit $l$ and an infinite limit.

(b) *If $f(x) \to \infty$ as $x \to \infty$ then $f(x) \not\to -\infty$ as $x \to \infty$* If $f(x) \to \infty$ as $x \to \infty$ then, using Definition 4.1.1 (with the special case $A = 1$), we see that there is some $X_1$ (with $X_1 > a$) such that

$$f(x) > 1$$

for all $x > X_1$. Hence $f(x) \nrightarrow -\infty$ as $x \to \infty$.

Hence $f(x)$ cannot have both $\infty$ and $-\infty$ as limits.

(c) *$f(x)$ cannot tend to two different finite limits* Suppose $f(x) \to l$ as $x \to \infty$ and $f(x) \to m$ as $x \to \infty$. Then either $l = m$ or $l \neq m$.

*Case $l \neq m$* If $l \neq m$, there is no loss of generality in assuming $l > m$. We now use Definition 4.1.3 (with the special case $\varepsilon = \frac{1}{2}(l - m)$). Since $f(x) \to l$ as $x \to \infty$ and $f(x) \to m$ as $x \to \infty$, there is some $X_1$ (with $X_1 > a$) such that

$$|f(x) - l| < \tfrac{1}{2}(l - m) \qquad (x > X_1), \tag{1}$$

$$|f(x) - m| < \tfrac{1}{2}(l - m) \qquad (x > X_1). \tag{2}$$

From relation (1) we have

$$l - \tfrac{1}{2}(l - m) < f(x) < l + \tfrac{1}{2}(l - m) \qquad (x > X_1),$$

i.e.

$$\tfrac{1}{2}(l + m) < f(x) < \tfrac{1}{2}(3l - m) \qquad (x > X_1). \tag{3}$$

Similarly, relation (2) reduces to

$$\tfrac{1}{2}(3m - l) < f(x) < \tfrac{1}{2}(l + m) \qquad (x > X_1). \tag{4}$$

Now (3) and (4) cannot both be satisfied for all $x > X_1$, since $f(x) > \frac{1}{2}(l + m)$ from (3) and $f(x) < \frac{1}{2}(l + m)$ from (4). Thus the assumption $l \neq m$ leads to a contradiction and must therefore be false. We are then left only with the possibility $l = m$, i.e. the two limits are in fact the same.

This completes the proof.

Since a function cannot have more than one limit, the limit, if it exists, is unique.

The familar 'sandwich rule' also has an analogue for functions.

THEOREM 4.1.4 Suppose $f, g, h$ are three functions such that $f(x)$, $g(x)$, $h(x)$ are all defined for $x > a$ and satisfy the inequality

$$g(x) \leqslant f(x) \leqslant h(x)$$

for all $x > a$. If $g(x) \to l$ as $x \to \infty$ and $h(x) \to l$ as $x \to \infty$, then

$$f(x) \to l \text{ as } x \to \infty.$$

*Proof* Let $\varepsilon > 0$. Since $g(x) \to l$ as $x \to \infty$ and $h(x) \to l$ as $x \to \infty$, there is some $X_1$ (with $X_1 > a$) such that

$$l - \varepsilon < g(x) < l + \varepsilon \tag{1}$$

$$l - \varepsilon < h(x) < l + \varepsilon \tag{2}$$

for all $x > X_1$. Hence, for all $x > X_1$,

$$l - \varepsilon < g(x) \leqslant f(x) \leqslant h(x) < l + \varepsilon.$$

And therefore $f(x) \to l$ as $x \to \infty$.

By this stage, the perceptive reader may be beginning to suspect that there are definite similarities between limits for sequences and limits for functions. Can these similarities be exploited to cut down on the number of proofs? Fortunately the answer is yes. The next result demonstrates the relation between limits for sequences and limits for functions. The underlying ideas are intuitively obvious, as recourse to a graph quickly illustrates. Suppose we have a function $f$ such that $f(x) \to l$ as $x \to \infty$. Figure 4.8 shows one such function. Now choose numbers $x_1, x_2, x_3$, etc., such that $x_n \to \infty$ as $n \to \infty$ and consider the sequence $(f(x_n))_{n=1}^{\infty}$ of real numbers. Then it would appear reasonable to guess that $f(x_n) \to l$ as $n \to \infty$. This establishes a link between the limit of $f(x)$ and the limit of the sequence $(f(x_n))$. Our result is contained in the next theorem.

THEOREM 4.1.5 Suppose that $f(x)$ is defined for all $x > a$, where $a$ is some given real number. Then $f(x) \to l$ as $x \to \infty$ if and only if $f(x_n) \to l$ as $n \to \infty$ for all sequences $(x_n)_{n=1}^{\infty}$ such that $x_n > a$ for all $n \geqslant 1$ and $x_n \to \infty$ as $n \to \infty$.

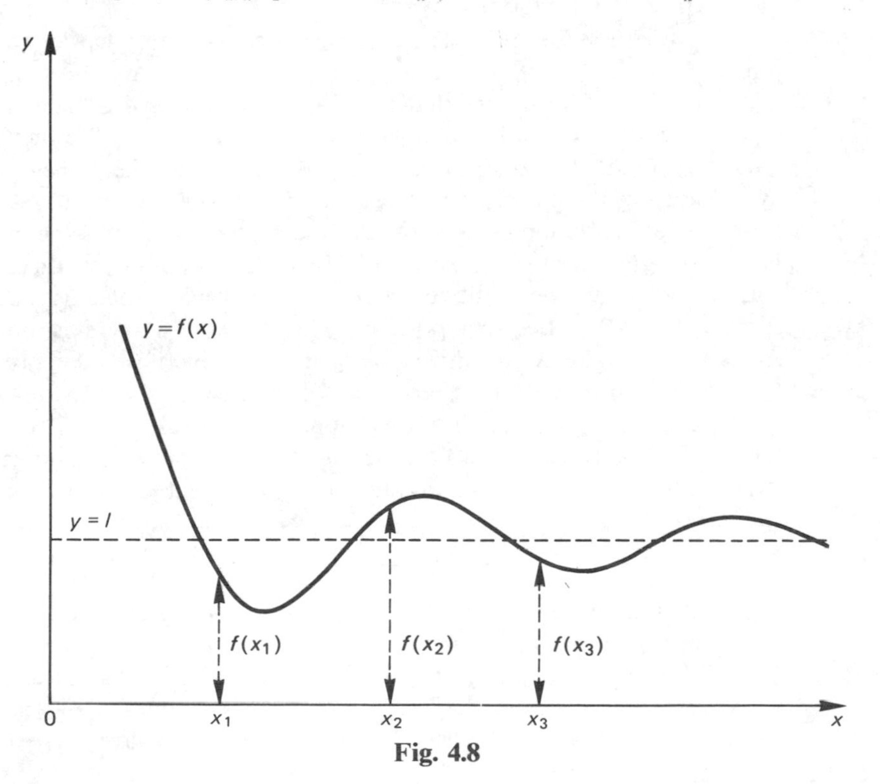


Fig. 4.8

*Proof* Since the theorem states that $f(x) \to l$ as $x \to \infty$ if and only if a certain condition is satisfied, we have two implications to prove. We must show that $f(x) \to l$ as $x \to \infty$ implies that all suitable sequences behave in the required way. Conversely we must show that the condition on sequences is sufficient to guarantee that $f(x) \to l$ as $x \to \infty$. The proof is, therefore, broken into two separate parts. One part deals with the 'only if' from 'if and only if'. The other concerns the 'if' in the double implication.

(a) '*Only if*' (In this part we assume that $f(x) \to l$ as $x \to \infty$ and prove that this implies $f(x_n) \to l$ as $n \to \infty$.)
Let $(x_n)_{n=1}^{\infty}$ be any sequence of real numbers such that $x_n > a$ for all $n \geqslant 1$ and $x_n \to \infty$ as $n \to \infty$, and let $\varepsilon > 0$. Since $f(x) \to l$ as $x \to \infty$, there is some $X_1$ (with $X_1 > a$) such that

$$|f(x) - l| < \varepsilon \tag{1}$$

for all $x > X_1$. Since $x_n \to \infty$ as $n \to \infty$, there is a positive integer $N$ such that

$$x_n > X_1 \tag{2}$$

for all $n > N$. Thus, for all $n > N$, $x_n > X_1$ and

$$|f(x_n) - l| < \varepsilon$$

by (1). Hence $f(x_n) \to l$ as $n \to \infty$.

This proves the first part. The converse is considerably harder to prove.

(b) '*If*' (In this part we assume that $f(x_n) \to l$ as $n \to \infty$ for all sequences $(x_n)_{n=1}^{\infty}$ such that $x_n > a$ for all $n \geqslant 1$ and $x_n \to \infty$ as $n \to \infty$. We then have to prove that this guarantees that $f(x) \to l$ as $x \to \infty$.) Let us begin by considering the consequences of assuming $f(x) \not\to l$ as $x \to \infty$. A close look at Definition 4.1.3 helps with the problem of expressing in symbols the statement $f(x) \not\to l$ as $x \to \infty$. Since $f(x) \not\to l$ as $x \to \infty$, there must be at least one positive value of $\varepsilon$ for which there is no corresponding $X_1$. Let this positive value be $\varepsilon_0$. As there is no corresponding value $X_1$ for this value $\varepsilon_0$, it means that if we take any number $X_1$ (however large) there is still a number $x > a$ such that $x > X_1$ and $|f(x) - l| \geqslant \varepsilon_0$. Since this happens for all values of $X_1$, we can use $X_1 = n$, where $n$ is a positive integer. Thus for each positive integer $n$ there is a corresponding number $x_n$ such that $x_n > a$, $x_n > n$ and

$$|f(x_n) - l| \geqslant \varepsilon_0. \tag{3}$$

From relation (3) we see that $f(x_n) \not\to l$ as $n \to \infty$. However, $(x_n)_{n=1}^{\infty}$ is a sequence such that $x_n > a$ and $x_n \to \infty$ as $n \to \infty$ (because $x_n > n$).

We have, therefore, proved that if $f(x) \not\to l$ as $x \to \infty$, then there is a sequence $(x_n)_{n=1}^{\infty}$ such that $x_n > a$, $x_n \to \infty$ as $n \to \infty$ and $f(x_n) \not\to l$ as $n \to \infty$. If no such sequence exists then it is impossible to have $f(x) \not\to l$

as $x\to\infty$. This means that if $f(x_n)\to l$ as $n\to\infty$ for every sequence $(x_n)_{n=1}^{\infty}$ such that $x_n > a$ $(n\geqslant 1)$ and $x_n\to\infty$ as $n\to\infty$, then we must also have $f(x)\to l$ as $x\to\infty$. The sequential condition is therefore sufficient to guarantee that $f(x)\to l$ as $x\to\infty$.

This completes the proof.

Similar results hold for the cases $f(x)\to\infty$ as $x\to\infty$ and $f(x)\to-\infty$ as $x\to\infty$. The theorems are stated without proof. The reader who wishes to check their veracity can produce a proof using the above arguments with appropriate modifications to suit the particular case.

THEOREMS 4.1.6 Suppose that $f(x)$ is defined for all $x > a$, where $a$ is some given real number. Then $f(x)\to\infty$ as $x\to\infty$ if and only if $f(x_n)\to\infty$ as $n\to\infty$ for all sequences $(x_n)_{n=1}^{\infty}$ such that $x_n > a$ for all $n\geqslant 1$ and $x_n\to\infty$ as $n\to\infty$.

THEOREM 4.1.7 Suppose that $f(x)$ is defined for all $x > a$, where $a$ is some given real number. Then $f(x)\to-\infty$ as $x\to\infty$ if and only if $f(x_n)\to-\infty$ as $n\to\infty$ for all sequences $(x_n)_{n=1}^{\infty}$ such that $x_n > a$ for all $n\geqslant 1$ and $x_n\to\infty$ as $n\to\infty$.

The connection between limits of sequences and limits of functions allows us to adapt theorems for sequences to produce their analogues for functions without the need for additional proof. For example, Theorem 2.2.5 concerning the algebra of limits immediately gives the corresponding result for functions.

THEOREM 4.1.8 (Algebra of Limits) Let $f(x)$, $g(x)$ be defined for all $x > a$, where $a$ is some given real number. Suppose also that $f(x)\to l$ as $x\to\infty$ and $g(x)\to m$ as $x\to\infty$. Then

(a) $f(x)+g(x)\to l+m$ as $x\to\infty$,
(b) $f(x)g(x)\to lm$ as $x\to\infty$.

If, in addition, $g(x)$ satisfies the condition $g(x)\neq 0$ for all $x > a$ and if $m\neq 0$, then

(c) $\dfrac{f(x)}{g(x)}\to\dfrac{l}{m}$ as $x\to\infty$.

*Proof* The proof follows immediately from Theorems 2.2.5 and 4.1.5. As an example of how these two results are used we look briefly at (c).

First use Theorem 4.1.5. It guarantees that for every sequence $(x_n)_{n=1}^{\infty}$ such that $x_n > a$ $(n\geqslant 1)$ and $x_n\to\infty$ as $n\to\infty$

$$f(x_n) \to l \qquad \text{as } n \to \infty,$$

$$g(x_n) \to m \qquad \text{as } n \to \infty,$$

since $f(x) \to l$ as $x \to \infty$ and $g(x) \to m$ as $x \to \infty$. Moreover, we are given that $m \neq 0$ and $g(x_n) \neq 0$. By Theorem 2.2.5, we see that

$$\frac{f(x_n)}{g(x_n)} \to \frac{l}{m}$$

as $n \to \infty$.

Since this is true for all sequences such that $x_n > a$, and $x_n \to \infty$, as $n \to \infty$ it follows from Theorem 4.1.5 that

$$\frac{f(x)}{g(x)} \to \frac{l}{m}$$

as $n \to \infty$.

The proofs of (a) and (b) are also rather obvious consequences of the results in Theorem 2.2.5 and Theorem 4.1.5.

Naturally, a proof of Theorem 4.1.8 is possible without recourse to sequences at all. For example, the first proof of the theorem could be proved in the following way. Let $\varepsilon > 0$. Since $f(x) \to l$ as $x \to \infty$ and $g(x) \to m$ as $x \to \infty$, there is some $X_1$ (with $X_1 > a$) such that

$$l - \varepsilon/2 < f(x) < l + \varepsilon/2 \qquad \text{for all } x > X_1$$

and

$$m - \varepsilon/2 < g(x) < m + \varepsilon/2 \qquad \text{for all } x > X_1.$$

By addition,

$$l + m - \varepsilon < f(x) + g(x) < l + m + \varepsilon$$

for all $x > X_1$ and so $f(x) + g(x) \to l + m$ as $x \to \infty$. This form of proof is modelled on the proof of Theorem 2.2.5 with the obvious modifications to allow for the fact that we are now dealing with functions and not sequences. The reader may feel that the direct proof of (a) (without recourse to sequences) is relatively painless, which is true. However, the advantages of using sequences becomes abundantly clear when the rule for the quotient $f(x)/g(x)$ is considered. In this case the direct proof contains many messy details.

Theorem 2.2.7 and its corollary also have obvious analogues for functions.

THEOREM 4.1.9 Suppose that $f(x)$ and $g(x)$ are defined for all $x > a$ and satisfy the inequality

$$f(x) \leqslant g(x) \qquad (x > a),$$

where $a$ is some given real number. If $f(x) \to l$ as $x \to \infty$ and $g(x) \to m$ as $x \to \infty$, then

$$l \leqslant m.$$

The proof of this result is left as an exercise for the reader. It demonstrates that weak inequalities are preserved when limits are taken. In the case of sequences, strict inequalities did not obey this rule exactly and the wary reader may suspect that the same is true for limits of functions. Such caution is exemplary! Even if the condition $f(x) \leqslant g(x)$ for all $x > a$ is replaced by the condition $f(x) < g(x)$ $(x > a)$, the conclusion remains in the form $l \leqslant m$ and that is the best we can do. For example, let $f(x) = 1/x^2$ and $g(x) = 1/x$ for $x > 1$. Then clearly $f(x) < g(x)$ for all $x > 1$, but

$$\lim_{x\to\infty} f(x) = 0 = \lim_{x\to\infty} g(x).$$

Thus the limits are equal despite the strictly inequality $f(x) < g(x)$ for $x > 1$.

In any particular situation the reader now has two different ways of attacking the solution of a problem concerning limits of functions. It is possible to use $\varepsilon$ and $X$, and tackle the problem directly. Alternatively, the link between sequences and limits of functions can be exploited and the problem resolved using sequences. Some problems are amenable to one approach and some to the other. The reader should be prepared to use either method, and will learn by experience which gives the neater solution to certain types of problems. Some examples are given below.

## *Examples 4.1.5*

1. Let $X$ be the set of all positive real numbers, i.e. $X = \{x \in \mathbb{R} : x > 0\}$. Let $f: X \to \mathbb{R}$ be defined in the following way:

$$f(x) = \begin{cases} 0 & \text{if } x \in X \text{ and } x \text{ is irrational,} \\ 1/q & \text{if } x \in X \text{ and } x \text{ is rational with } x = p/q, \\ & \text{where } p, q \text{ are positive integers with no} \\ & \text{common factor greater than 1.} \end{cases}$$

(The condition that $p, q$ have no common factor exceeding 1 means that the fraction $p/q$ is in its lowest terms, i.e. common factors of numerator and denominator have been cancelled out.)

This example looks formidably complicated, but there is no need to be intimidated. It is really quite easy if we use sequences.

From Theorems 4.1.5 to 4.1.7 we notice that if $f(x)$ tends to any limit (finite or infinite) as $x \to \infty$, then $f(x_n)$ must also tend to this limit as $n \to \infty$ for all suitable sequences. Thus if we find two sequences $(x_n)_{n=1}^{\infty}$ and $(y_n)_{n=1}^{\infty}$ such that $x_n \to \infty$ as $n \to \infty$ and $y_n \to \infty$ as $n \to \infty$ and $f(x_n)$ and $f(y_n)$ tend to two **different** limits as $n \to \infty$, then $f(x)$ cannot tend to any limit as $x \to \infty$. In this case we choose such sequences, one consisting entirely of suitable rational numbers and the other consisting of irrational numbers. Let us use

$$x_n = \frac{2n-1}{2} \qquad (n = 1, 2, 3, 4, \ldots),$$

$$y_n = n\sqrt{2} \qquad (n=1,2,3,4,\dots).$$

Clearly $x_n \to \infty$ as $n \to \infty$, $y_n \to \infty$ as $n \to \infty$ and $x_n > 0$, $y_n > 0$. Now $(x_n)_{n=1}^{\infty}$ is the sequence $\frac{1}{2}, \frac{3}{2}, \frac{5}{2}, \frac{7}{2}, \dots$, and $f(x_n) = \frac{1}{2}$. By definition $f(y_n) = 0$. Hence $f(x_n) \to \frac{1}{2}$ as $n \to \infty$, and $f(y_n) \to 0$ as $n \to \infty$. The fact that

$$\lim_{n\to\infty} f(x_n) \neq \lim_{n\to\infty} f(y_n)$$

guarantees that $f(x)$ does not tend to any limit as $x \to \infty$.

The above example illustrates one type of problem for which the sequential criterion is particularly powerful. It can be used to great effect to prove that a function $f$ does not have a limit. For it is then sufficient to find two sequences $(x_n)$ and $(y_n)$ (satisfying appropriate conditions) such that $f(x_n)$ and $f(y_n)$ do not have the same limit.

On the other hand if we want to prove that a function does possess a limit, then the $\varepsilon$, $X$ criterion is frequently more useful, since the sequential criterion requires a certain condition to be proved for **all** sequences.

Of course, it would be possible to formulate a solution for Example 1 using $\varepsilon$ and $X$, but the reader will find it difficult to express the explanation clearly and convincingly using this approach.

2. Show that $$\frac{x^2-1}{x^2+1} \to 1 \qquad \text{as } x \to \infty.$$

*Solution* For $x > 0$

$$\frac{x^2-1}{x^2+1} = \frac{1-1/x^2}{1+1/x^2}. \tag{1}$$

As $x \to \infty$, $1/x^2 \to 0$ and so by the algebra of limits

$$1 - 1/x^2 \to 1 - 0 = 1$$

$$1 + 1/x^2 \to 1 + 0 = 1$$

as $x \to \infty$. Hence, by equation (1),

$$\frac{x^2-1}{x^2+1} \to \frac{1}{1} = 1$$

as $x \to \infty$.

3. For all $x > 0$

$$\frac{1-1/x}{1+1/x^2} = \frac{x^2-x}{x^2+1} \leqslant \frac{x^2+x\sin x}{x^2+1} \leqslant \frac{x^2+x}{x^2+1} = \frac{1+1/x}{1+1/x^2}.$$

As $x \to \infty$,

$$\frac{1-1/x}{1+1/x^2} \to \frac{1-0}{1+0} = 1,$$

$$\frac{1+1/x}{1+1/x^2} \to \frac{1+0}{1+0} = 1$$

by the algebra of limits. Using the sandwich rule we see that

$$\frac{x^2 + x\sin x}{x^2+1} \to 1$$

as $x \to \infty$.

4. For all $x > 1$,

$$\frac{x^3 + x\sin x}{x^2+1} \geqslant \frac{x^3 - x}{x^2+1} = x\left(\frac{x^2-1}{x^2+1}\right).$$

Now
$$\frac{x^2-1}{x^2+1} = 1 - \frac{2}{x^2+1} > 1 - \frac{2}{5} = \frac{3}{5}$$

for all $x > 2$. Thus for $x > 2$

$$\frac{x^3 + x\sin x}{x^2+1} > \frac{3}{5}x.$$

Now $3x/5 \to \infty$ as $x \to \infty$ and, by Theorem 4.1.1, $(x^3 + x\sin x)/(x^2+1) \to \infty$ as $x \to \infty$.

---

## *EXERCISES 4.1.4*

**1** For each of the following functions $f$, decide whether $f(x)$ tends to a limit as $x \to \infty$. When the limit exists find it.

(a) $f(x) = \dfrac{2x^2+1}{3x^2+3x+1}$; (b) $f(x) = \dfrac{1-\sqrt{x}}{1+\sqrt{x}}$;

(c) $f(x) = \dfrac{x^3}{x^2+1}$; (d) $f(x) = \dfrac{x^3 \sin x}{x^2+1}$;

(e) $f(x) = \dfrac{x^3 \sin^2 x}{x^2+1}$; (f) $f(x) = \dfrac{x^2 \sin^2 x}{x^2+1}$;

(g) $f(x) = \dfrac{x\sin x}{x^2+1}$; (h) $f(x) = \sinh x$;

(i) $f(x) = \sqrt{x^2+1} - \sqrt{x^2-1}$.

In each case justify your answer.

**2** Suppose $f(x)$ and $g(x)$ are defined for all $x > a$, where $a$ is some given real number. Suppose also that $f(x) \to \infty$ as $x \to \infty$ and $|g(x)| \leqslant k$ for all $x > a$, where $k$ is some positive real number. Show that $f(x)+g(x)\to\infty$ as $x\to\infty$. Give examples to show that (a) $f(x)g(x)$ need not tend to any limit as $x\to\infty$, (b) if $f(x)g(x)$ does tend to a limit as $x\to\infty$, then this limit can be infinite or it can be finite.

**3** Show that if $f(x)\to l$ and $g(x)\to\infty$ as $x\to\infty$, then $f(x)+g(x)\to\infty$ as $x\to\infty$.

**4** Give examples of functions $f$ and $g$ to illustrate each of the following possibilities.

(a) $f(x)\to\infty$ and $g(x)\to-\infty$ as $x\to\infty$, but $f(x)+g(x)$ does not tend to any limit as $x\to\infty$.
(b) $f(x)\to\infty$ and $g(x)\to-\infty$ as $x\to\infty$, but $f(x)+g(x)\to-\infty$ as $x\to\infty$.
(c) $f(x)\to\infty$ and $g(x)\to-\infty$ as $x\to-\infty$, but $f(x)+g(x)\to l$ as $x\to\infty$, where $l$ is some given real number.

The beginning of this section was devoted to limits as $x\to\infty$ and the material was presented in a very leisurely manner. Naturally, a corresponding theory exists for limits as $x\to-\infty$. By now, the reader should have sufficient knowledge and experience to be able to envisage how to frame the definitions and to decide which results will be useful in actually finding limits. The following definitions and theorems will, therefore, be given without further explanations and proofs. They are all, in fact, rather obvious analogues of the ones we have already met.

DEFINITION 4.1.4 Suppose that $f(x)$ is defined for all $x<a$, where $a$ is some given real number. (*Note:* The phrase $f(x)$ is defined for all $x<a$ means that the domain of definition of $f$ contains the interval $(-\infty,a)$.) We say that $f(x)\to\infty$ as $x\to-\infty$ if given any $A>0$ (however large) there exists a corresponding $X_1$ (with $X_1<a$) such that $f(x)>A$ for **all** $x<X_1$.

DEFINITION 4.1.5 Suppose that $f(x)$ is defined for all $x<a$, where $a$ is some given real number. We say that $f(x)\to l$ as $x\to-\infty$ if given any $\varepsilon>0$ (however small) there exists a corresponding $X_1$ (with $X_1<a$) such that

$$|f(x)-l|<\varepsilon$$

for **all** $x<X_1$.

DEFINITION 4.1.6 Suppose that $f(x)$ is defined for all $x<a$, where $a$ is some given real number. We say that $f(x)\to-\infty$ as $x\to-\infty$ if given $A>0$

(however large) there exists a corresponding $X_1$ (with $X_1 < a$) such that $f(x) < -A$ for **all** $x < X_1$.

As in the case of limits as $x \to \infty$, we find that the limit (if it exists) is unique.

THEOREM 4.1.10 Suppose that $f(x)$ is defined for all $x < a$, where $a$ is some given real number. Then $f(x)$ cannot tend to more than one limit as $x \to -\infty$.

Theorems 4.1.2 to 4.1.4, which contain practical tools for establishing the value of limits, have their obvious analogues.

THEOREM 4.1.11 Suppose that $f(x)$ and $g(x)$ are defined for all $x < a$ and satisfy the inequality $f(x) \geqslant g(x)$ for all $x < a$, where $a$ is some given real number. If $g(x) \to \infty$ as $x \to -\infty$, then $f(x) \to \infty$ as $x \to -\infty$.

THEOREM 4.1.12 (Sandwich Rule) Suppose that $f(x)$, $g(x)$, $h(x)$ are all defined for all $x < a$ and satisfy the inequality

$$g(x) \leqslant f(x) \leqslant h(x)$$

for all $x < a$, where $a$ is some given real number. If $g(x) \to l$ as $x \to -\infty$ and $h(x) \to l$ as $x \to -\infty$, then $f(x) \to l$ as $x \to -\infty$.

THEOREM 4.1.13 Suppose that $f(x)$, $g(x)$ are defined for all $x < a$ and satisfy the inequality

$$f(x) \leqslant g(x)$$

for all $x < a$, where $a$ is some given real number. If $g(x) \to -\infty$ as $x \to -\infty$ then $f(x) \to -\infty$ as $x \to -\infty$.

In the case of limits as $x \to -\infty$, it is again possible to write down results demonstrating the link between sequences and limits of functions. As before, there are rules for limits of sums, products etc.

THEOREM 4.1.14 Suppose that $f(x)$, $g(x)$ are defined for all $x < a$, where $a$ is some given real number. If $f(x) \to l$, $g(x) \to m$ as $x \to -\infty$, then

(a) $f(x) + g(x) \to l + m$ as $x \to -\infty$;
(b) $f(x)g(x) \to lm$ as $x \to -\infty$.

Moreover, if $f$ satisfies the additional conditions that $g(x) \neq 0$ for $x < a$ and $m \neq 0$, then

(c) $\dfrac{f(x)}{g(x)} \to \dfrac{l}{m}$ as $x \to -\infty$.

THEOREM 4.1.15 Suppose that $f(x)$, $g(x)$ are defined for all $x < a$, and satisfy the inequality

$$f(x) \leqslant g(x) \qquad (x < a),$$

where $a$ is some given real number. If $f(x) \to l$ and $g(x) \to m$ as $x \to -\infty$, then

$$l \leqslant m.$$

Let us now return to one of the functions sketched at the beginning of the section. In Fig. 4.3 we have the graph of $y = 1/x$. It is easy to see that $1/x \to 0$ as $x \to \infty$ and also that $1/x \to 0$ as $x \to -\infty$. But what happens if $x \to 0$?

The graph of $y = 1/x$ is in two separate pieces and if we start with a positive value for $x$ and let $x$ approach zero on the right-hand portion (i.e. on the piece in the first quadrant), then it looks as if $y \to \infty$, i.e. if we let $x$ approach zero in such a way that it only takes positive values, then it seems that $y \to \infty$. On the other hand if we let $x$ approach zero on the left-hand piece of the graph (i.e. the piece in the third quadrant for which $x < 0$), then it looks as if $y \to -\infty$. So in this case it looks as if we have a reasonably sensible notation of a limit provided $x$ approaches zero from just one side. If we want to use symbols to express the fact that $x$ is approaching zero from the right (i.e. only through values of $x$ greater than 0) then we write either $\lim_{x \to 0+}$ or we use the phrase $1/x \to \infty$ as $x \to 0+$. This limit is called a one-sided limit because we only approach zero from the right-hand side. In general, if we are interested in a one-sided limit of $f(x)$ as $x$ approaches a point $a$ from the right, then we use $\lim_{x \to a+} f(x)$ or we discuss what happens to $f(x)$ as $x \to a+$. The phrase $x \to a+$ expresses the fact that we are only considering what happens to $f(x)$ as $x$ approaches $a$ through values which exceed $a$.

Naturally, we also have one-sided limits from the left. Suppose we wish to investigate what happens to $f(x)$ as $x$ approaches a point $a$ through values less than $a$. For such limits we use either $\lim_{x \to a-}$ or we discuss what happens to $f(x)$ as $x \to a-$, the symbols $a-$ denoting the fact that we are only considering what happens to $f(x)$ as we approach $a$ through values less than $a$ (i.e. we approach $a$ from the left).

One-sided limits can be useful when there is a nasty break in the graph of a function or when the function is defined only at one side of the point in question. For example $1/x$ has a nasty break at 0, but we can write $1/x \to \infty$ as $x \to 0+$ and $1/x \to -\infty$ as $x \to 0-$. The natural logarithm $\log x$, however, is defined only for $x > 0$ and the reader can check by drawing the graph that it looks as if $\log x \to -\infty$ as $x \to 0+$.

In both the examples above, we notice that the function was not defined at the point itself, for neither $1/x$ nor $\log x$ have a value when $x = 0$. All that we required in taking the one-sided limits, $\lim_{x\to 0+}$ and $\lim_{x\to 0-}$, is that the function is defined at all points sufficiently close to 0, either on the right of 0 or on the left of 0, whichever is appropriate. In taking limits we totally ignore what happens at the point; we only make statements about what happens as $x$ approaches the point. Thus, for example, when we define $\lim_{x\to a+} f(x)$ we will need to stipulate that $f(x)$ is defined for $a < x < a + R$, where $R$ is some positive real number, i.e. we will need to know that the domain of definition of $f$ contains some interval $(a, a + R)$ for some $R > 0$. This guarantees that $f(x)$ has a value when $x$ is sufficiently close to $a$, but on the right of $a$. Once the one-sided limits have been defined, they can be used to define the two-sided limit. The same theorems are needed for the one-sided limits and the two-sided limits. We will therefore write down all the definitions and then simply state results like the sandwich rule and the algebra of limits once. The reader can make appropriate minor adjustments to adapt them for use with one-sided or two-sided limits. The definitions follow a similar pattern to that of those we have already met, with one significant change.

To pinpoint the exact place at which the alteration must be made, let us return to the example $f(x) = 1/x$ $(x \neq 0)$. We have already suggested that $1/x \to \infty$ as $x \to 0+$. In fact, given any positive real number $A$ (however large) we notice that $f(x) > A$ for all $x$ such that $0 < x < 1/A$, i.e. $f(x) > A$ for all positive values of $x$ sufficiently close to zero. If we write $\delta = 1/A$, then $\delta > 0$ and in this example $f(x) > A$ for all $x$ such that $0 < x < \delta$ (see Fig. 4.9). Thus the conditions we need are similar to those in Definition 4.1.1 except that we now have $0 < x < \delta$ instead of $x > X_1$. For we are now interested in values of $x$ close to 0, whereas Definition 4.1.1 was concerned with all sufficiently large values of $x$.

In general, if we are considering what happens to $f(x)$ as $x \to a+$, then we will be interested in determining whether certain conditions are satisfied for all $x$ sufficiently close to $a$, but greater than $a$, i.e. we will need to know whether the criteria are satisfied for all $x$ such that $a < x < a + \delta$ where $\delta$ is some positive real number. We are now in a position to formulate the definitions.

DEFINITION 4.1.7 Suppose that $f(x)$ is defined for all $x$ such that $a < x < a + R$, where $R$ is some given positive real number. Then we say that $f(x) \to \infty$ as $x \to a+$, if given any $A > 0$ (however large) there is a corresponding $\delta > 0$ (with $\delta \leqslant R$) such that $f(x) > A$ for **all** $x$ such that $a < x < a + \delta$.

DEFINITION 4.1.8 Suppose $f(x)$ is defined for all $x$ such that $a - R < x < a$, where $R$ is some given positive real number. Then we say that $f(x) \to \infty$ as

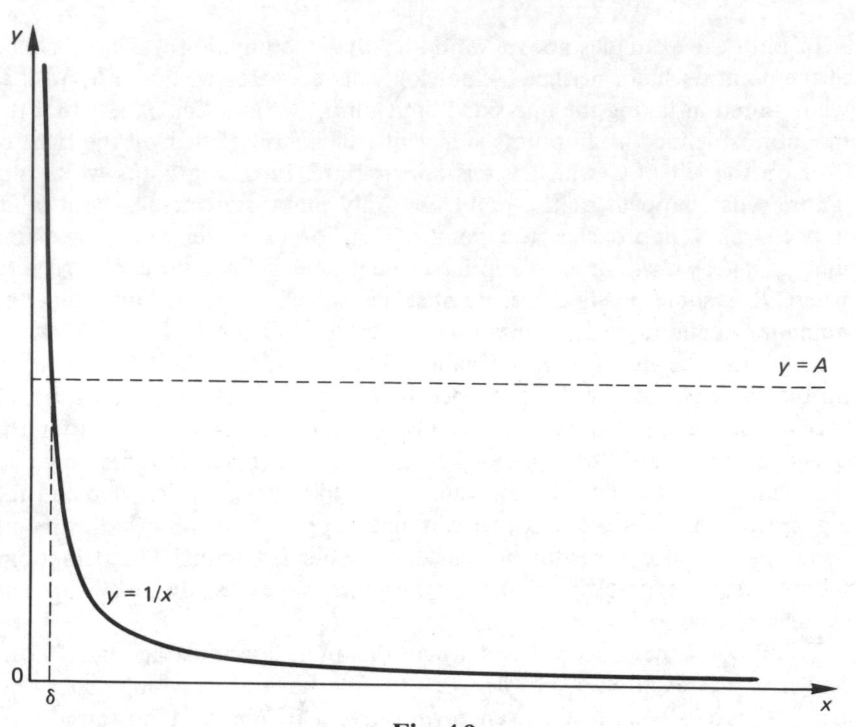


Fig. 4.9

$x \to a-$, if given any $A > 0$ (however large) there is a corresponding $\delta > 0$ (with $\delta \leqslant R$) such that $f(x) > A$ for **all** $x$ such that $a - \delta < x < a$.

DEFINITION 4.1.9 Suppose $f(x)$ is defined for all $x$ such that $a < x < a + R$, where $R$ is some given positive real number. Then we say that $f(x) \to l$ as $x \to a+$ if given any $\varepsilon > 0$ (however small) there is a corresponding $\delta > 0$ (with $\delta \geqslant R$) such that

$$|f(x) - l| < \varepsilon$$

for **all** $x$ such that $a < x < a + \delta$.

DEFINITION 4.1.10 Suppose $f(x)$ is defined for all $x$ such that $a - R < x < a$, where $R$ is some given positive real number. Then we say that $f(x) \to l$ as $x \to a-$ if given any $\varepsilon > 0$ (however small) there is a corresponding $\delta > 0$ (with $\delta \leqslant R$) such that

$$|f(x) - l| < \varepsilon$$

for **all** $x$ such that $a - \delta < x < a$.

DEFINITION 4.1.11 Suppose $f(x)$ is defined for all $x$ such that $a<x<a+R$, where $R$ is some given positive real number. Then we say that $f(x)\to -\infty$ as $x\to a+$ if, given any $A>0$ (however large), there is a corresponding $\delta>0$ (with $\delta\leqslant R$) such that $f(x)<-A$ for **all** $x$ such that $a<x<a+\delta$.

DEFINITION 4.1.12 Suppose $f(x)$ is defined for all $x$ such that $a-R<x<a$, where $R$ is some given positive real number. Then we say that $f(x)\to -\infty$ as $x\to a-$ if, given any $A>0$ (however large), there is a corresponding $\delta>0$ (with $\delta\leqslant R$) such that $f(x)<-A$ for **all** $x$ such that $a-\delta<x<a$.

The two-sided limit, $\lim_{x\to a}f(x)$, exists if $\lim_{x\to a+}f(x)$ and $\lim_{x\to a-}f(x)$ both exist and are the same. This common limit is called $\lim_{x\to a}f(x)$. By combining the previous definitions, it is easy to formulate definitions for the two-sided limits. One case, in particular, merits mention as it turns out to be extremely important in practice. This special case is the one in which $\lim_{x\to a+}f(x)$ and $\lim_{x\to a-}f(x)$ are finite and equal. The relevant definition is given below. Since $\lim_{x\to a+}f(x)$ and $\lim_{x\to a-}f(x)$ both exist, $f(x)$ must be defined for all $x$ such that $a<x<a+R$ and also for all $x$ such that $a-R<x<a$. The first condition gives $0<x-a<R$ and the second condition is $-R<x-a<0$. These two can be combined to give the inequality $0<|x-a|<R$.

DEFINITION 4.1.13 Suppose $f(x)$ is defined for all $x$ such that $0<|x-a|<R$, where $R$ is some given positive real number. Then we say that $f(x)\to l$ as $x\to a$, if, given any $\varepsilon>0$ (however small), there is a corresponding $\delta>0$ (with $\delta\leqslant R$) such that

$$|f(x)-l|<\varepsilon$$

for **all** $x$ such that $0<|x-a|<\delta$.

(*Note:* This definition is couched in the same terms as Heine's definition which dates back to over a century ago. The one superficial difference is that Heine used the Greek letter $\eta$ in place of the letter $\delta$ we use today.)

We notice from the inequality $0<|x-a|<\delta$ that the point $x=a$ is specifically excluded. Thus the statement $f(x)\to l$ as $x\to a$ tells us how $f(x)$ behaves as $x$ approaches $a$. However, it gives us absolutely no information about what happens at $a$ itself. As before, it can be proved that a function cannot have more than one limit.

Two results are of great practical importance as they are frequently used in determining limits. They are analogues of results which we have met already.

THEOREM 4.1.16 (Sandwich Rule) Suppose $f(x)$, $g(x)$, $h(x)$ are defined and satisfy the inequality

$$g(x) \leqslant f(x) \leqslant h(x)$$

for all $x$ such that $0<|x-a|<R$, where $R$ is some given positive real number. If $g(x) \to l$ as $x \to a$ and $h(x) \to l$ as $x \to a$, then $f(x) \to l$ as $x \to a$.

THEOREM 4.1.17 (Algebra of Limits) Suppose $f(x)$, $g(x)$ are defined for all $x$ such that $0<|x-a|<R$, where $R$ is some given positive real number. If $f(x) \to l$ as $x \to a$ and $g(x) \to m$ as $x \to a$, then

(a) $f(x)+g(x) \to l+m$ as $x \to a$;
(b) $f(x)g(x) \to lm$ as $x \to a$.

If in addition $g(x)$ satisfies the condition $g(x) \neq 0$ for all $x$ such that $0<|x-a|<R$ and if $m \neq 0$, then

(c) $\dfrac{f(x)}{g(x)} \to \dfrac{l}{m}$ as $x \to a$.

THEOREM 4.1.18 Suppose $f(x)$, $g(x)$ are defined for all $x$ such that $0<|x-a|<R$ and satisfy the inequality

$$f(x) \leqslant g(x) \qquad (0<|x-a|<R),$$

where $R$ is some given positive real number. If $f(x) \to l$ and $g(x) \to m$ as $x \to a$, then

$$l \leqslant m.$$

Proofs of these results can easily be constructed along the same lines as those given for earlier theorems of the same form. These results are not restricted to the case of limits as $x \to a$; they can be used for limits as $x \to a+$ and limits as $x \to a-$, provided that the wording is very slightly modified (in the obvious way) to make it appropriate for the case being considered.

As the reader may expect, Theorems 4.1.1 and 4.1.2 have obvious analogues for limits as $x \to a$, etc. However, these analogues for infinite limits are less frequently used than the results for finite limits.

We have already noticed that the link between sequences and limits of functions can be of practical use in some circumstances. In the case in which the limit as $x \to a$ is being considered we have the following statement.

THEOREM 4.1.19 Suppose $f(x)$ is defined for all $x$ such that $0<|x-a|<R$, where $R$ is some given positive real number. Then $f(x) \to l$ as $x \to a$ if and only if $f(x_n) \to l$ as $n \to \infty$ for all sequences $(x_n)_{n=1}^{\infty}$ such that $0<|x_n-a|<R$ (for all $n \geqslant 1$) and $x_n \to a$ as $n \to \infty$.

Let us now see if we can put these theorems to good use by working some examples.

## *Examples 4.1.6*

1. The function $f: \mathbb{R} \to \mathbb{R}$ is defined by

$$f(x) = \begin{cases} x & (x < 0), \\ 1 & (x = 0), \\ 1 + x & (0 < x < 1), \\ 3 & (x = 1), \\ 2x^2 & (x > 1). \end{cases}$$

Decide whether the following limits exist and find their value when they do exist:

(a) $\lim_{x \to 0-} f(x)$; (b) $\lim_{x \to 0+} f(x)$; (c) $\lim_{x \to 0} f(x)$; (d) $\lim_{x \to 1} f(x)$.

Solution

(a) If $x < 0$, $f(x) = x$. As $x \to 0-$, $f(x) \to 0$. Hence

$$\lim_{x \to 0-} f(x) = 0.$$

(b) If $0 < x < 1$, $f(x) = 1 + x$. We see that $1 + x \to 1$ as $x \to 0+$ (using the algebra of limits). Hence

$$\lim_{x \to 0+} f(x) = 1.$$

(c) Since

$$\lim_{x \to 0-} f(x) \neq \lim_{x \to 0+} f(x),$$

we see that $\lim_{x \to 0} f(x)$ does not exist.

(d) If $0 < x < 1$, $f(x) = 1 + x$. We see that $f(x) \to 2$ as $x \to 1-$.

i.e.
$$\lim_{x \to 1-} f(x) = 2.$$

If $1 < x$, $f(x) = 2x^2$. We see that $f(x) \to 2$ as $x \to 1+$.

Hence
$$\lim_{x \to 1+} f(x) = 2 = \lim_{x \to 1-} f(x).$$

Thus $\lim_{x \to 1} f(x) = 2$.
The fact that $f(1) = 3$ has no bearing at all on whether $\lim_{x \to 1} f(x)$ exists or not. In deciding whether the limit exists or not we are interested in what happens as $x$ approaches 1. We ignore what happens at $x = 1$.

2. Let
$$f(x) = \frac{x \sin x}{x^2 + 1}$$

for all $x \in \mathbb{R}$. For all $x$,

$$|f(x)| \leqslant \frac{|x|}{x^2 + 1},$$

i.e.
$$\frac{-|x|}{x^2+1} \leqslant f(x) \leqslant \frac{|x|}{x^2+1}.$$

Using the algebra of limits, we see that
$$\frac{-|x|}{x^2+1} \to 0, \frac{|x|}{x^2+1} \to 0 \text{ as } x \to 0.$$

By the sandwich rule, therefore,
$$f(x) \to 0 \text{ as } x \to 0.$$

In this example a slightly modified version of the sandwich rule can be used. Suppose that
$$|f(x)| \leqslant |g(x)| \tag{1}$$
for all $x$ such that $0 < |x-a| < R$, where $R$ is some positive real number. Suppose also that $g(x) \to 0$ as $x \to a$. Then we can deduce from inequality (1) that $f(x) \to 0$ as $x \to a$. For, by definition, given any $\varepsilon > 0$, there is a corresponding $\delta > 0$ (with $\delta \leqslant R$) such that $|g(x)| < \varepsilon$ for all $x$ such that $0 < |x-a| < \delta$ because $g(x) \to 0$ as $x \to a$. Hence, for $0 < |x-a| < \delta$,
$$|f(x)| \leqslant |g(x)| < \varepsilon$$
and, therefore, $f(x) \to 0$ as $x \to a$.

3. Let $f$ be defined on $\mathbb{R}\backslash\{0\}$ by
$$f(x) = x\sin(\tfrac{1}{x}) \qquad (x \neq 0).$$
For $x \neq 0$ we have
$$|f(x)| = |x\sin(\tfrac{1}{x})| \leqslant |x|,$$
and therefore $f(x) \to 0$ as $x \to 0$.
The graph of $y = x\sin(1/x)$ is shown as Fig. 4.10.

4. Let $f$ be defined on $\mathbb{R}\backslash\{0\}$ by
$$f(x) = \sin(1/x) \qquad (x \neq 0).$$
Then, using sequences, we can show that $f(x)$ does not tend to any limit as $x \to 0$. We prove this by choosing two sequences $(x_n)_{n=1}^{\infty}$ and $(y_n)_{n=1}^{\infty}$ so that $x_n \to 0$, $y_n \to 0$ as $n \to \infty$, but
$$\lim_{n\to\infty} f(x_n) \neq \lim_{n\to\infty} f(y_n).$$
To do this choose
$$x_n = 1/n\pi \qquad (n = 1, 2, 3, 4, \ldots)$$
$$y_n = \frac{2}{(4n+1)\pi} \qquad (n = 1, 2, 3, 4, \ldots).$$

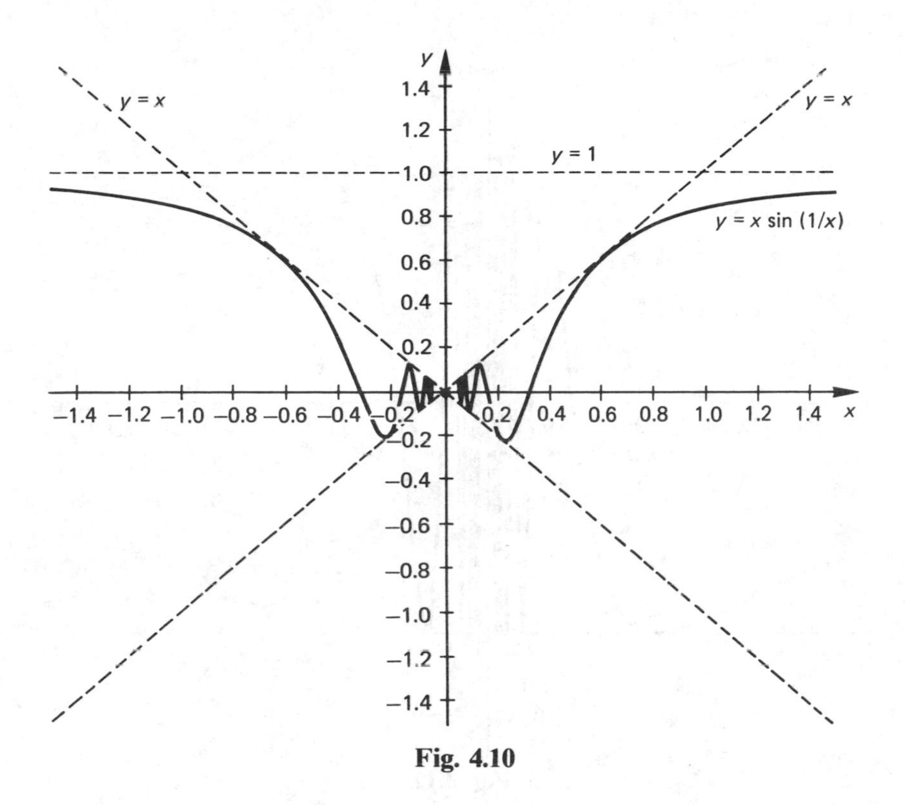


**Fig. 4.10**

Clearly $x_n \to 0$ and $y_n \to 0$ as $n \to \infty$.

Moreover, $\qquad f(x_n) = \sin(1/x_n) = \sin n\pi = 0$

$$f(y_n) = \sin(1/y_n) = \sin[(4n+1)\pi/2] = 1.$$

Hence $f(x_n) \to 0$ as $n \to \infty$ and $f(y_n) \to 1$ as $n \to \infty$. Since the two limits are different, $f(x)$ does not tend to any limit as $x \to 0$. Since $x_n > 0$ and $y_n > 0$ for all $n$, we see that $f(x)$ does not tend to any limit as $x \to 0+$. The graph is shown as Fig. 4.11.

Our final example in this section concerns a limit which has rather important implications, as we will see later.

5. Let $C$ be a circle centre $O$ and radius 1. Suppose the arc $AB$ of the circle $C$ subtends an angle $x$ radians at $O$, where $0 < x < \pi/2$. Suppose also that the tangent to the circle $C$ at the point $A$ cuts $OB$ extended at $Q$, and that $P$ is the point on $OB$ such that $AP$ is perpendicular to $OB$. Using elementary trigonometry we see that the length of $AP$ is $\sin x$ and the length of $AQ$ is $\tan x$. Now the area of a triangle is $\frac{1}{2}$ base × height. Since $OB$ is of length 1, the triangle $OBA$ has area $\frac{1}{2}\sin x$. The sector $OAB$ of the circle has area $\frac{1}{2}x$ (since $x$ is measured in radians) and the triangle $OAQ$ has area $\frac{1}{2}\tan x$. Using Fig. 4.12 we can see that

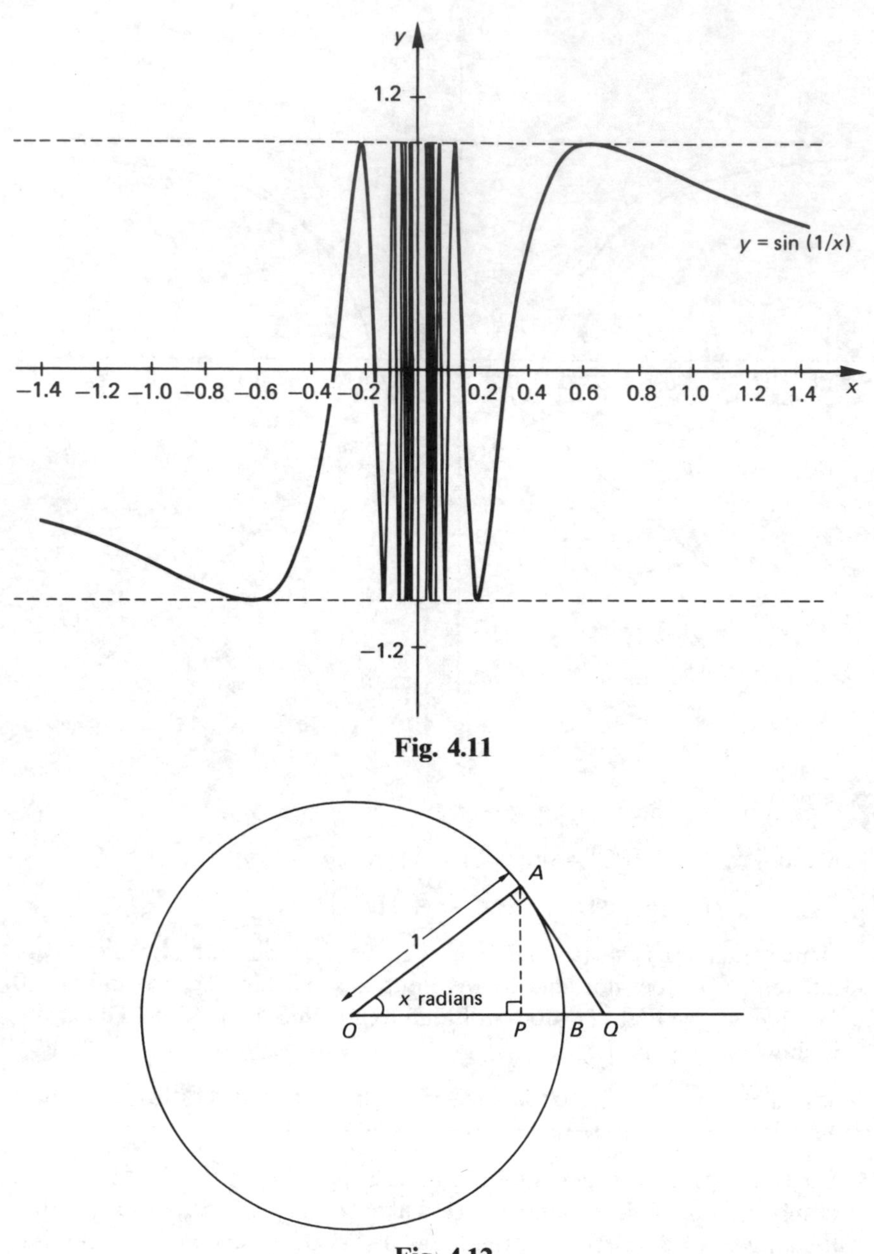


**Fig. 4.11**

**Fig. 4.12**

$$\text{area of } \triangle OBA < \text{area of sector } OBA < \text{area } \triangle OQA,$$

i.e.
$$\tfrac{1}{2}\sin x < \tfrac{1}{2}x < \tfrac{1}{2}\tan x.$$

Thus, for $0 < x < \pi/2$,
$$\sin x < x < \tan x. \tag{1}$$

Now if $0 < x < \pi/2$, then $0 < x/2 < \pi/4 < \pi/2$ and, therefore,

$$0 < \sin(x/2) < x/2$$

by (1). Hence for $0 < x < \pi/2$,

$$\cos x = 1 - 2\sin^2(x/2) > 1 - 2(x/2)^2 = 1 - \tfrac{1}{2}x^2, \tag{2}$$

and therefore, from (1),

$$1 - \tfrac{1}{2}x^2 < \cos x < \frac{\sin x}{x} < 1 \tag{3}$$

for $0 < x < \pi/2$. Since $\cos(-x) = \cos x$ and $\sin(-x) = -\sin x$ we see that (3) is also true if $-\pi/2 < x < 0$. Thus if $0 < |x| < \pi/2$, then

$$1 - \tfrac{1}{2}x^2 < \cos x < \frac{\sin x}{x} < 1.$$

Using the sandwich rule we see that

$$\frac{\sin x}{x} \to 1 \text{ as } x \to 0$$

and also $\cos x \to 1$ as $x \to 0$. The limit for $(\sin x)/x$ will be used later when we come to the question of differentiation of trigonometric functions.

Incidently, this limit also helps to clear up a point which arose in connection with Example 3. The graph appears to indicate that $x(\sin 1/x) \to 1$ as $x \to \infty$, which we can easily verify. For if $x > 0$, then,

$$x \sin\left(\frac{1}{x}\right) = \frac{\sin\left(\frac{1}{x}\right)}{\frac{1}{x}}.$$

As $x \to \infty$, $1/x \to 0$ and therefore

$$\frac{\sin(1/x)}{(1/x)} \to 1.$$

i.e. $$x \sin(1/x) \to 1 \text{ as } x \to \infty.$$

Similarly, $x \sin(1/x) \to 1$ as $x \to -\infty$.

## *EXERCISES 4.1.5*

**1** For each of the following functions $f$ and points $a$, decide whether $f(x)$ tends to a limit as $x \to a$. When the limit exists find it.

(a) $f(x) = \dfrac{2x^2 + 1}{3x^2 + 3x + 1}$, $\quad a = 0$;

(b) $f(x)=\dfrac{2x^2+1}{3x^2+3x+1}$, $\quad a=1$;

(c) $f(x)=\dfrac{(x^2+1)\sin x}{x}$ $\quad (x\neq 0),\ a=0$;

(d) $f(x)=\dfrac{x\sin(1/x)}{x^2+1}$ $\quad (x\neq 0),\ a=0$;

(e) $f(x)=\sin\left[x+\left(\dfrac{1}{x}\right)\right]$ $\quad (x\neq 0),\ a=0$;

(f) $f(x)=\dfrac{1}{\sqrt{x^2+1}-\sqrt{x^2-1}}$ $\quad (x\neq 1),\ a=0$;

(g) $f(x)=\dfrac{\sin \alpha x}{x}$ $\quad (x\neq 0),\ a=0$, where $\alpha$ is any real number;

(h) $f(x)=\dfrac{\sin x}{x^2}$ $\quad (x\neq 0),\ a=0$;

(i) $f(x)=\begin{cases} x \ (x \text{ rational}) \\ x^2 \ (x \text{ irrational})\end{cases}$, $\quad a=0$;

(j) $f(x)=\begin{cases} 0 \ (x \text{ rational}) \\ 1 \ (\text{irrational})\end{cases}$, $\quad a=1$.

In each case give reasons to justify your answer.

**2** Let $f:\mathbb{R}\to\mathbb{R}$ be defined by

$$f(x)=\begin{cases} \sin x & (x<0), \\ 1 & (x=0), \\ x^2 & (0<x<1), \\ 2x-1 & (1\leqslant x). \end{cases}$$

Decide whether $\lim_{x\to 0}f(x)$ and $\lim_{x\to 1}f(x)$ exist. Find any limit which exists.

**3** Give examples of functions $f$ and $g$ to illustrate each of the following possibilities:

(a) $f(x)\to 0$ as $x\to 1$, $g(x)\to 0$ as $x\to 1$, but $f(x)/g(x)$ does not tend to any limit as $x\to 1$.

(b) $f(x)\to 0$ as $x\to 1$, $g(x)\to 0$ as $x\to 1$ and $f(x)/g(x)$ tends to $\infty$ as $x\to 1$.

(c) $f(x)\to 0$ as $x\to 1$, $g(x)\to 0$ as $x\to 1$ and $f(x)/g(x)$ tends to 3 as $x\to 1$.

**4** Show that for all real numbers $x$

$$|\sin x|\leqslant |x|.$$

By using the formula

$$\cos x - \cos a = 2\sin\left(\frac{x+a}{2}\right)\sin\left(\frac{a-x}{2}\right),$$

prove that $$|\cos x - \cos a| \leqslant |x-a|.$$

Deduce that $\cos x \to \cos a$ as $x \to a$.

Use the same method to show that $\sin x \to \sin a$ as $x \to a$. (*Warning:* Remember that $x$ is measured in radians).

## 4.2 CONTINUITY

In the sixth form, students frequently think of a continuous function as one whose graph can be sketched without taking the pencil off the paper. Their idea of a continuous function, therefore, is one which has no breaks. Suppose we examine this notion in a little more detail. A function $f$ does not have a break at a point $a$ if $f(x) \to f(a)$ as $x \to a$. This is precisely the condition we need for continuity. Since limits are defined in terms of symbols, the requirement $f(x) \to f(a)$ as $x \to a$ does not depend on existence of a graphical representation of $f$.

DEFINITION 4.2.1 Suppose that $f(x)$ is defined for all $x$ such that $a - R < x < a + R$, where $R$ is some given positive real number. Then we say that **$f$ is continuous at** $a$ if $f(x) \to f(a)$ as $x \to a$, i.e. if given any $\varepsilon > 0$, there is a corresponding $\delta > 0$ (with $\delta \leqslant R$) such that

$$|f(x) - f(a)| < \varepsilon$$

for all $x$ such that $|x - a| < \delta$.

We notice that our definiton of continuity of $f$ at a point $a$ is subject to the condition that $f(x)$ is defined for all $x$ such that $a - R < x < a + R$. This means that the domain of definition of the function $f$ contains some interval of the form $(a - R, a + R)$ $(R > 0)$. It will be tactily assumed that such a condition is satisfied whenever the phrase '$f$ is continuous at $a$' is used. (*Note:* This convention is not used universally. Frequently the restriction on the type of point is lifted and the $\varepsilon, \delta$ condition in the definition is replaced by the requirement that $|f(x) - f(a)| < \varepsilon$ for all $x$ in the domain of definition of $f$ for which $|x - a| < \delta$.) Extensions of these ideas to continuity at other types of points are possible, but we will not use them.

In analysis the concept of continuity on an interval is frequently needed.

DEFINITION 4.2.2 Let $a, b$ be two real numbers with $a < b$ and suppose that $f(x)$ is defined for all $x$ such that $a < x < b$. We say that $f$ is continuous on the open interval $(a, b)$ if $f$ is continuous at all points $c$ for which $a < c < b$.

DEFINITION 4.2.3 Let $a, b$ be two real numbers with $a < b$ and suppose that $f(x)$ is defined for all $x$ such that $a \leqslant x \leqslant b$. We say that $f$ **is continuous on the closed interval** $[a, b]$ if $f$ is continuous at all points $c$ such that $a < c < b$ and $\lim_{x \to a+} f(x) = f(a)$, $\lim_{x \to b-} f(x) = f(b)$.

The use of the one-sided limits at the end points $a, b$ ensures that we do not need to consider what happens outside the interval $[a, b]$. Instead of using the phrase '$f$ is continuous on the closed interval $[a, b]$', we sometimes just write $f: [a, b] \to \mathbb{R}$ is continuous.

As the reader might expect, there are results concerning sums, products, quotients, etc., of continuous functions. In preparation for the proof of the result for quotients we establish the following result.

THEOREM 4.2.1 Suppose that $f(x)$ is defined for all $x$ such that $a - R < x < a + R$, for some positive real number $R$. If $f$ is continuous at $a$ and $f(a) \neq 0$, then there is some $\delta > 0$, (with $\delta \leqslant R$) such that $f(x) \neq 0$ for all $x$ such that $a - \delta < x < a + \delta$.

*Proof* Since $f(a) \neq 0$, it follows that

$$|f(a)| > 0.$$

Now $f$ is continuous at $a$ and so using the definition of continuity with the special case $\varepsilon = |f(a)|$, we see that there is a $\delta > 0$ (with $\delta \leqslant R$) such that

$$|f(x) - f(a)| < |f(a)| \tag{1}$$

for all $x$ such that $a - \delta < x < a + \delta$. From (1) we see that $f(x) \neq 0$ for $a - \delta < x < a + \delta$.

We can now deal with sums, products and quotients of continuous functions.

THEOREM 4.2.2 Suppose that $f(x)$, $g(x)$ are defined for all $x$ such that $a - R < x < a + R$, for some positive real number $R$. If $f$, $g$ are continuous at $a$, then

(a) $f + g$ is continuous at $a$;
(b) $fg$ is continuous at $a$;
(c) $f/g$ is continuous at $a$, provided $g(a) \neq 0$.

*Proof* Since $f, g$ are continuous at $a$, $f(x) \to f(a)$ and $g(x) \to g(a)$ as $x \to a$. By Theorem 4.1.17, $f(x) + g(x) \to f(a) + g(a)$ as $x \to a$ and $f(x)g(x) \to f(a)g(a)$ as $x \to a$. This proves (a) and (b).

Since $g(a) \neq 0$, Theorem 4.2.1 ensures that there is some $\delta > 0$ (with $\delta \leqslant R$)

such that $g(x) \neq 0$ for all $x$ such that $a-\delta<x<a+\delta$. From Theorem 4.1.17 it now follows that

$$\frac{f(x)}{g(x)} \to \frac{f(a)}{g(a)}$$

as $x \to a$. Hence $f/g$ is continuous at $a$.

Similar results follow for functions continuous on an interval. For example we have the following:

THEOREM 4.2.3 Let $f\colon [a,b] \to \mathbb{R}$ be continuous and let $g\colon [a,b] \to \mathbb{R}$ be continuous. Then

(a) $f+g$ is continuous on $[a,b]$,
(b) $fg$ is continuous on $[a,b]$,
(c) $f/g$ is continuous at all points of $[a,b]$ at which $g$ is not zero.

Theorems about continuous functions are hardly very productive without a stock of continuous functions. Do we have such a collection available? From the definition (with $\delta=\varepsilon$) it is clear that the identity function $f$ given by $f(x)=x$ for all $x \in \mathbb{R}$ is continuous at all points of $\mathbb{R}$. Using the algebra of limits, it follows that all positive integral powers of $x$ are continuous at all points of $\mathbb{R}$, and therefore every polynomial is continuous at all points of $\mathbb{R}$. Again, an appeal to the algebra of limits assures us that the quotient of two polynomials (i.e. a rational function) is continuous at all points of $\mathbb{R}$ at which the denominator is not zero. The familiar trigonometric functions sine and cosine, the hyperbolic functions sinh and cosh, and the exponential function provide us with a further stock of functions which are continuous at every point of $\mathbb{R}$. Since these functions are usually formally defined in terms of power series, the reader is asked to accept on trust the statement that they are continuous. A rigorous detailed proof requires properties of power series and is beyond the scope of this volume. We can further enlarge our stock of continuous functions by using the theorem on inverse functions which appears later in this section. With the help of this result we can conclude that $\log x$ is continuous at all points $x$ for which $x>0$ and we can investigate where $n$th roots are continuous.

In examples we may need more than just sums, products and quotients of continuous functions. For example these rules, on their own, are not really adequate when we come to handle objects like $\sin(e^x)$ which is often referred to as a function of a function or a composite function—to use mathematical terminology. Let us recall how composite functions are defined. Suppose we have two functions $g\colon X \to Y$ and $f\colon Y \to Z$ (see Fig. 4.13). If we take $x \in X$, then $g(x) \in Y$ and then, using $f$, we see that $f(g(x)) \in Z$. The two together, therefore, associate with each $x \in X$ a unique element $z \in Z$, and so they define

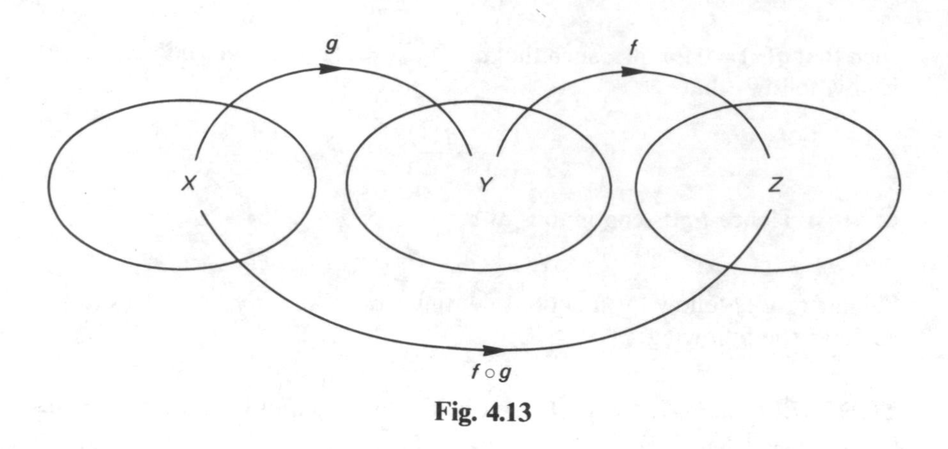


**Fig. 4.13**

a function from $X$ into $Z$. This function is denoted by $f \circ g$. It has the property that, for each $x \in X$, the image $(f \circ g)(x)$ is the element $f(g(x)) \in Z$.

*Examples 4.2.1*

1. Let $g: \mathbb{R} \to \mathbb{R}$ and $f: \mathbb{R} \to \mathbb{R}$ be defined by

$$g(x) = e^x \qquad (\text{all } x \in \mathbb{R}),$$
$$f(x) = \sin x \qquad (\text{all } x \in \mathbb{R}).$$

Then $(f \circ g): \mathbb{R} \to \mathbb{R}$ and $(g \circ f): \mathbb{R} \to \mathbb{R}$ are both defined and

$$(f \circ g)(x) = f(g(x)) = \sin(e^x)$$
$$(g \circ f)(x) = g(f(x)) = e^{\sin x}$$

for all $x \in \mathbb{R}$.

2. Let $Y$ be the set of all positive real numbers and suppose that $g: \mathbb{R} \to Y$ and $f: Y \to \mathbb{R}$ are defined by

$$g(x) = x^2 + 1 \qquad (\text{all } x \in \mathbb{R}),$$
$$f(y) = \log y \qquad (\text{all } y \in Y, \text{ i.e. all } y > 0)$$

Then $(f \circ g): \mathbb{R} \to \mathbb{R}$ is defined and

$$(f \circ g)(x) = f(g(x)) = \log(x^2 + 1)$$

for all $x \in \mathbb{R}$. The composite function $(g \circ f): Y \to Y$ is also defined and

$$(g \circ f)(y) = g(f(y)) = (\log y)^2 + 1$$

for all $y \in Y$, i.e. for all $y > 0$.

In the present context the question which exercises us is whether the continuity of a composite function can be deduced from the continuity of the separate functions. Let us take two functions $g: X \to Y$, and $f: Y \to Z$ where $X, Y, Z$ are all sets of real numbers. Suppose also that all the following conditions are satisfied:

(a) $X$ contains an interval of the form $(a-R, a+R)$ for some positive real number $R$;
(b) $g(a)=b$;
(c) $g$ is continuous at $a$;
(d) $Y$ contains an interval of the form $(b-s, b+s)$ for some positive real number $s$;
(e) $f$ is continuous at $b$.

Now let $\varepsilon>0$. Since $f$ is continuous at $b$, there is some $\delta_1>0$ (with $\delta_1 \leqslant s$) such that

$$|f(y)-f(b)|<\varepsilon \tag{1}$$

for all $y$ such that $b-\delta_1<y<b+\delta_1$.
Since $g$ is continuous at $a$, there is some $\delta>0$ (with $\delta \leqslant R$) such that

$$|g(x)-g(a)|<\delta_1 \tag{2}$$

for all $x$ such that $a-\delta<x<a+\delta$. Relation (2) can be written as

$$b-\delta_1<g(x)<b+\delta_1 \tag{3}$$

for all $x$ such that $a-\delta<x<a+\delta$, because $g(a)=b$. Hence, if $a-\delta<x<a+\delta$, then

$$|f(g(x))-f(b)|<\varepsilon,$$

i.e.
$$|f(g(x))-f(g(a))|<\varepsilon.$$

This means that given any $\varepsilon>0$, there exists a corresponding $\delta>0$ such that

$$|f(g(x))-f(g(a))|<\varepsilon$$

for all $x$ such that $a-\delta<x<a+\delta$. Hence $f \circ g$ is continuous at $a$.

We now record this result in the following theorem.

THEOREM 4.2.4 Let $f, g$ be two real valued functions such that

(a) $g$ is continuous at $a$;
(b) $g(a)=b$,
(c) $f$ is continuous at $b$.

Then the composite function $(f \circ g)$ given by

$$(f \circ g)(x)=f(g(x))$$

is continuous at $a$.

*Examples 4.2.2*

1. Let $f\colon \mathbb{R}\to\mathbb{R}$, $g\colon \mathbb{R}\to\mathbb{R}$ be given by

$$g(x)=e^x \qquad \text{(all real } x\text{)}$$

$$f(x)=\sin x \qquad \text{(all real } x\text{).}$$

Then $f$ is continuous at all points of $\mathbb{R}$ and $g$ is continuous at all points of $\mathbb{R}$.
Thus the composite functions $f\circ g$ and $g\circ f$ are both continuous at all points of $\mathbb{R}$, i.e. $\sin(e^x)$ and $e^{\sin x}$ are both continuous at all points of $\mathbb{R}$.

2. Let $h\colon \mathbb{R}\to\mathbb{R}$ be defined by

$$h(x)=\begin{cases}\sin(1/x) & x\neq 0,\\ 0 & x=0.\end{cases}$$

Write $f(x)=\sin x$ for all real $x$ and let $g(x)=1/x$ $(x\neq 0)$. Then $g$ is continuous at all points other than the origin, i.e. $g$ is continuous at all points of $\mathbb{R}\backslash\{0\}$. Moreover, $f$ is continuous at all points of $\mathbb{R}$. The composite function $f\circ g$ is therefore continuous at all points of $\mathbb{R}\backslash\{0\}$, i.e. $h$ is continuous at all points of $\mathbb{R}\backslash\{0\}$. Now $h(x)$ does not tend to any limit as $x\to 0$ (see Examples 4.1.6). Hence $h(x)\not\to h(0)$ as $x\to 0$ and $h$ is not continuous at the origin.

3. Let $h\colon \mathbb{R}\to\mathbb{R}$ be defined by

$$h(x)=\begin{cases}x\sin(1/x) & (x\neq 0),\\ 0 & (x=0).\end{cases}$$

We see from the previous example that $\sin(1/x)$ is continuous at all points of $\mathbb{R}\backslash\{0\}$. Using the result for the product of two continuous functions, we see that $x\sin(1/x)$ is continuous at all points of $\mathbb{R}\backslash\{0\}$, i.e. $h$ is continuous at all points of $\mathbb{R}\backslash\{0\}$. But what happens at the origin? Is $h$ continuous there? To decide what happens at the origin we consider $|h(x)-h(0)|$. If $x\neq 0$ then

$$\begin{aligned}|h(x)-h(0)|&=|x\sin(1/x)-0|\\&=|x\sin(1/x)|\leqslant|x|.\end{aligned}$$

Hence, $|h(x)-h(0)|\to 0$ as $x\to 0$, i.e. $h(x)\to h(0)$ as $x\to 0$ and $h$ is also continuous at the origin. Thus the function $h$ is continuous at all points of $\mathbb{R}$.

## *EXERCISES 4.2.1*

1 Use the results of questions in Exercises 4.1.5 to show that the sine and cosine are continuous at all points of $\mathbb{R}$.

**2** Decide whether each of the following functions $f$ is continuous at the point $a$ indicated.

(a)
$$f(x)=[x],\ a=0$$

where $[x]$ is the integer part of $x$, i.e. $[x]$ is the largest integer which does not exceed $x$.

(b) $f(x)=\begin{cases} x^2 & (x<0) \\ \sin x & (x\geqslant 0) \end{cases} \quad (a=0).$

(c) $f(x)=\begin{cases} \sqrt{|x|}\sin(1/x) & (x\neq 0) \\ 0 & (x=0) \end{cases} \quad (a=0).$

**3** The function $f\colon \mathbb{R}\to\mathbb{R}$ is defined by

$$f(x)=\begin{cases} x & \text{when } x \text{ is rational,} \\ x^2 & \text{when } x \text{ is irrational.} \end{cases}$$

Show that $f$ is continuous at 0 and continuous at 1, but discontinuous at all other points.

**4** The function $f\colon \mathbb{R}\to\mathbb{R}$ is defined by

$$f(x)=\begin{cases} 0 & \text{when } x \text{ is rational,} \\ 1 & \text{when } x \text{ is irrational.} \end{cases}$$

Show that $f$ is discontinuous everywhere.

**5** (a) The function $f\colon \mathbb{R}\to\mathbb{R}$ is defined by

$$f(x)=\begin{cases} x & \text{when } x \text{ is rational,} \\ 0 & \text{when } x \text{ is irrational.} \end{cases}$$

Show that $f$ is continuous at the origin and discontinuous at all other points.

(b) Construct a function $f\colon \mathbb{R}\to\mathbb{R}$ which is continuous at $\sqrt{2}$ and discontinuous at all other points.

## 4.3 SOME PROPERTIES OF CONTINUOUS FUNCTIONS ON CLOSED INTERVALS

In section 4.2 we saw how certain combinations of continuous functions are themselves continuous. For example, the product of two continuous functions is continuous. So we have had plenty of practice in deciding whether a function is continuous or not. But what interesting properties do these continuous functions possess anyway?

Suppose we take a continuous function whose graph is fairly simple to draw and consider its behaviour on a closed interval. It is hard to imagine that a function continuous on a closed interval could be unbounded on this interval. Perhaps that is hardly surprising as we can prove that a function continuous on a closed is bounded on that interval. Again, our result needs to be proved using symbols as it must cover the case of a continuous function whose graph cannot be drawn.

THEOREM 4.3.1 Suppose $f(x)$ is defined for all $x$ such that $a \leqslant x \leqslant b$ and $f$ is continuous on the closed interval $[a, b]$. Then $f$ is bounded on $[a, b]$, i.e. there is some positive real number $k$ such that $|f(x)| \leqslant k$ for all $x \in [a, b]$.

*Proof* Let us see what would happen if $f$ is not bounded on $[a, b]$. In this case, if we choose any positive integer $n$, then there is some point $x_n$ with $a \leqslant x_n \leqslant b$ such that

$$|f(x_n)| > n. \tag{1}$$

The points of the sequence $(x_n)_{n=1}^{\infty}$ all satisfy the inequality $a \leqslant x_n \leqslant b$, i.e. it is bounded. It must, therefore, have a convergent subsequence $(x_{n_k})_{k=1}^{\infty}$ by the corollary to Theorem 2.5.4. Let

$$c = \lim_{k \to \infty} x_{n_k}.$$

Then $a \leqslant c \leqslant b$ since $a \leqslant x_n \leqslant b$ for all $n$. Since $x_{n_k} \to c$ as $k \to \infty$ and $f$ is continuous on $[a, b]$ it follows that

$$f(x_{n_k}) \to f(c) \tag{2}$$

as $k \to \infty$. However, condition (1) shows that

$$|f(x_{n_k})| > n_k \geqslant k$$

and so
$$|f(x_{n_k})| \to \infty \tag{3}$$

as $k \to \infty$. This contradicts relation (2). Hence the assumption that $f$ is not bounded on $[a, b]$ leads to a contradiction and this assumption must be false. It therefore follows that $f$ is bounded on $[a, b]$ and the proof is complete.

In the statement of the theorem we deliberately included the phrase 'the closed interval $[a, b]$' because the result is not true for open intervals, which may suprise the reader considerably. For example, if $f(x) = 1/x$ $(x > 0)$ then $f$ is continuous on the open interval $(0, 1)$, but $f$ is not bounded on this interval. In fact, given any $A > 0$ (however large) we see that $f(x) > A$ for all $x$ such that $0 < x < 1/A$.

Now suppose we consider the set $S$ of all values assumed by $f(x)$ when $a \leqslant x \leqslant b$. If $f$ is continuous on $[a, b]$, then $S$ is bounded. Since it is bounded above it has a supremum and similarly it has an infimum. Are there values of $x$ for which $f(x)$ has the same value as this supremum? The next result answers this question.

THEOREM 4.3.2 Suppose $f(x)$ is defined for all $x$ such that $a \leqslant x \leqslant b$ and suppose also that $f$ is continuous on the closed interval $[a, b]$. Let

$$M = \sup_{a \leqslant x \leqslant b} f(x), \qquad m = \inf_{a \leqslant x \leqslant b} f(x).$$

Then there are points $c, d$ with $a \leqslant c, d \leqslant b$ such that $f(c) = M, f(d) = m$. (This means that $M$ is the largest value of $f(x)$ on $[a, b]$ and $m$ is the smallest value.)

*Proof* We begin by considering what happens if the result is not true. Since

$$M = \sup_{a \leqslant x \leqslant b} f(x),$$

it follows that $f(x) \leqslant M$ for all $x \in [a, b]$. If there is no point $c \in [a, b]$ for which $f(c) = M$, then $f(x) < M$ for all $x \in [a, b]$. Hence $M - f(x) \neq 0$ for all $x \in [a, b]$. Using the continuity of $f$ and the algebra of limits we see that the function $g$ given by

$$g(x) = \frac{1}{M - f(x)} \qquad (a \leqslant x \leqslant b)$$

is continuous on $[a, b]$. By Theorem 4.3.1, $g$ must therefore be bounded on $[a, b]$ and so there is some positive real number $k$ such that $|g(x)| \leqslant k$ for all $x \in [a, b]$.

i.e.
$$\frac{1}{M - f(x)} = \left|\frac{1}{M - f(x)}\right| \leqslant k \qquad (a \leqslant x \leqslant b),$$

which gives
$$f(x) \leqslant M - \frac{1}{k}$$

for all $x \in [a, b]$. Hence $M - 1/k$ is an upper bound for $f$ on $[a, b]$ and it is less than the supremum $M$. Thus the assumption that there is no $c$ for which $f(c) = M$ leads to a contradiction. This assumption is therefore false and there is some point $c \in [a, b]$ for which $f(c) = M$. Similarly, there is a point $d \in [a, b]$ for which $f(d) = m$. This completes the proof.

The fact that there are points $c, d \in [a, b]$ at which $f(c) = M$ and $f(d) = m$ means that the values $M, m$ are actually attained by $f(x)$ on $[a, b]$. For this reason Theorems 4.3.1 and 4.3.2 are often taken together and stated succinctly in the following way. A function which is continuous on a closed interval is bounded and attains its bounds.

What other properties do we expect a continuous function to possess? Let us look at $\sin x$. We know that $\sin(\pi/2) = 1$ and $\sin(3\pi/2) = -1$, and we always take it for granted that $\sin x$ takes every value between 1 and $-1$ for $\pi/2 \leqslant x \leqslant 3\pi/2$, because $\sin x$ is continuous. In general, if the function $f$ is continuous on $[a, b]$, then we draw the graph $y = f(x)$ so that $y$ takes every value between $f(a)$ and $f(b)$ for $a \leqslant x \leqslant b$. This property is captured in the intermediate value theorem.

THEOREM 4.3.3 (Intermediate Value Theorem) Suppose $f(x)$ is defined for all $x$ such that $a \leqslant x \leqslant b$. Suppose also that $f$ is continuous on the closed interval $[a,b]$, that $f(a) \neq f(b)$ and that $K$ is any number strictly between $f(a)$ and $f(b)$. Then there is some $c$ such that $a < c < b$ and $f(c) = K$. (i.e. $f(x)$ assumes every value between $f(a)$ and $f(b)$ at least once on $[a,b]$.)

*Proof* Since $f(a) \neq \mathrm{f}(b)$, we begin by assuming $f(a) < f(b)$. Let $K$ be any number such that

$$f(a) < K < f(b).$$

(The proof that the value $K$ is not omitted hinges on the fact that the real numbers obey the completeness axiom and so there are no gaps in the real numbers. Since the completeness axiom is vital, it is not suprising that the proof contains essentially two sets of real numbers belonging to $[a,b]$—the set for which $f(x) < K$ and the set for which $f(x) \geqslant K$. As $x$ crosses from one set to the other, $f(x)$ assumes the value $K$.)

Let $S$ be the set of all real numbers $x$ such that $a \leqslant x \leqslant b$ and $f(x) < K$. Then the set $S$ is bounded above and it is not empty as $a \in S$. By the completeness axiom $S$ has a supremum. Let $c = \sup S$. Then

$$a \leqslant c \leqslant b \tag{1}$$

since $S$ only contains real numbers $x$ for which $a \leqslant x \leqslant b$. We now need to prove $c > a$ and $c < b$.

(a) $c > a$ Since $f$ is continuous on $[a,b]$, $f(x) \to f(a)$ as $x \to a+$. Using the definition of a limit (with the special value $\varepsilon = K - f(a)$), we see that there is some $\delta > 0$ (with $\delta \leqslant b - a$) such that

$$|f(x) - f(a)| < K - f(a)$$

for all $x$ such that $a \leqslant x < a + \delta$. This inequality gives

$$f(x) < K$$

for all $x$ such that $a \leqslant x < a + \delta$. In particular $f(a + \delta/2) < K$ and hence $a + \delta/2 \in S$, which implies that $c \geqslant a + \delta/2$, i.e. $c > a$.

(b) $c < b$ Since $f$ is continuous on $[a,b]$, $f(x) \to f(b) > K$ as $x \to b-$. From this we can deduce that there is some $\delta_1 > 0$ (with $\delta_1 \leqslant b - a$) such that $f(x) \geqslant K$ for all $x$ such that $b - \delta_1 < x \leqslant b$ and these values of $x$ do not belong to $S$. Thus all numbers $x$ in $S$ satisfy the inequality $a \leqslant x \leqslant b - \delta_1 < b$ and therefore $c = \sup S \leqslant b - \delta_1 < b$.

Our results together imply that

$$a < c < b. \tag{2}$$

To complete the proof, all that now remains is to demonstrate that $c$ has the required property, viz. that $f(c) = K$. To do this we show that $f(c)$ satisfies both the inequalities $f(c) \leqslant K$ and $f(c) \geqslant K$.

(c) $f(c) \leqslant K$ By definition $c = \sup S$ and so for every positive integer $n$ there is a point $x_n \in [a, b]$ such that $c - 1/n < x_n \leqslant c$ and $x_n \in S$. Thus

$$f(x_n) < K$$

because $x_n \in S$. Now $f$ is continuous at $c$ and $x_n \to c$ as $n \to \infty$. Hence

$$f(c) = \lim_{n \to \infty} f(x_n) \leqslant K, \tag{3}$$

using the sequential criterion.

(d) $f(c) \geqslant K$ If $b \geqslant x > c = \sup S$, then $x \notin S$ and $f(x) \geqslant K$. Since $f$ is continuous at $c$,

$$f(c) = \lim_{x \to c+} f(x) \geqslant K. \tag{4}$$

From relations (3) and (4) we see that $f(c) = K$ and the proof is complete.

The reader can readily see how the proof needs to be modified for a function for which $f(a) > f(b)$.

Theorems 4.3.3 and 4.3.2 together show that if $f$ is continuous on $[a, b]$, then $f(x)$ assumes every value between $M$ and $m$ at least once on $[a, b]$, where $M = \sup_{a \leqslant x \leqslant b} f(x)$ and $m = \inf_{a \leqslant x \leqslant b} f(x)$. If we happen to be considering an example in which $f(x)$ takes both positive and negative values on $[a, b]$, then $m < 0$ and $M > 0$. The intermediate value theorem then allows us to deduce that there is a point $c \in [a, b]$ at which $f(c) = 0$. This idea lies behind many of the examples on continuous functions given to undergraduates.

## *Examples 4.3.1*

1. Show that the equation $xe^{\sin x} = \cos x$ has a solution in the interval $(0, \pi/2)$.

*Solution* Whenever trigonometric functions are used in analysis it is **always** assumed (unless the contrary is specifically stated) that the unit of measure is a **radian**. In this particular example, therefore, the quantity $x$ is measured in radians.

The problem is solved by introducing a function $f$ given by

$$f(x) = xe^{\sin x} - \cos x.$$

Using the rules for composite functions, and products and differences of continuous functions, we see that $f$ is continuous at all points of $\mathbb{R}$. In particular, it is continuous on the closed interval $[0, \pi/2]$.

Moreover,

$$f(0) = -1,$$
$$f(\pi/2) = \pi e/2 > 0,$$

and, therefore, by the intermediate value theorem (with $K = 0$) there is a point $x_0$ with $0 < x_0 < \pi/2$ such that $f(x_0) = 0$, i.e. $x = x_0$ is a solution of the given equation.

2. Suppose $f(x)$ is defined for all $x$ such that $a \leqslant x \leqslant b$ and suppose also that $a \leqslant f(x) \leqslant b$ whenever $a \leqslant x \leqslant b$. Show that if $f$ is continuous on $[a, b]$, then the equation

$$f(x) = x$$

has a root in $[a, b]$.

*Solution* Define a function $g$ by

$$g(x) = f(x) - x \qquad (a \leqslant x \leqslant b).$$

Then $g$ is continuous on $[a, b]$. Moreover,

$$g(a) = f(a) - a \geqslant 0$$
$$g(b) = f(b) - b \leqslant 0.$$

If $f(a) = a$, then $x = a$ is a root of the given equation. If $f(b) = b$, then $x = b$ is a root of the given equation. If $f(a) \neq a$ and $f(b) \neq b$, then

$$g(a) = f(a) - a > 0,$$
$$g(b) = f(b) - b < 0,$$

and there is a number $c$ with $a < c < b$ such that $g(c) = 0$ by the intermediate value theorem,

i.e.

$$f(c) = c.$$

Thus in all cases the equation $f(x) = x$ has a root in $[a, b]$.

---

When the function $f$ satisfies the conditions given in Example 2 above, it may be possible to locate a root of $f(x) = x$ by using successive approximations. As a first approximation choose a number $x_1$ with $a \leqslant x_1 \leqslant b$. Define the sequence $(x_n)_{n=1}^{\infty}$ by the recurrence relation

$$x_{n+1} = f(x_n) \qquad (n = 1, 2, 3, \ldots).$$

Then $a \leqslant x_n \leqslant b$ for all $n$, since $a \leqslant f(x) \leqslant b$ whenever $a \leqslant x \leqslant b$. If the sequence $(x_n)_{n=1}^{\infty}$ converges to a point $d$, then $a \leqslant d \leqslant b$ and $f(x_n) \to f(d)$ as $n \to \infty$, because $f$ is continuous on $[a, b]$. However, $f(x_n) = x_{n+1}$ and $x_{n+1} \to d$ as $n \to \infty$, and therefore $f(x_n) \to d$ as $n \to \infty$. Since a sequence cannot have more than one limit we must have $f(d) = d$. Thus if $(x_n)$ converges, its limit $d$ satisfies the equation $f(d) = d$. So $x = d$ is a root of $f(x) = x$, if $(x_n)$ converges.

Of course, there is still an element of doubt, as we do not know whether the sequence converges. We overcome this difficulty by using the intermediate value theorem.

Let us look at a particular example. Let $f(x) = \cos x$, $a = 0$ and $b = 1$. Then $f$ satisfies all the required conditions since $f$ is continuous on $[0, 1]$ and $0 \leqslant \cos x \leqslant 1$ when $0 \leqslant x \leqslant 1$. We therefore have a root of the equation $\cos x = x$ in the interval $[0, 1]$.

Now let us see that happens if we use successive approximations. We will use $x_1 = 0.5$ and

$$x_{n+1} = \cos x_n \qquad (n = 1, 2, 3, \ldots).$$

Using a pocket calculator we have
$x_2 = 0.878$, $x_3 = 0.639$, $x_4 = 0.803$, $x_5 = 0.695$, ..., $x_{10} = 0.745$, $x_{11} = 0.735$, $x_{12} = 0.742$, $x_{13} = 0.737$, $x_{14} = 0.740$, .... (The reader is reminded that we are using radians not degrees for measuring $x$.) It looks as if the sequence $(x_n)$ is converging to 0.74 (correct to two decimal places) and it is reasonable to guess that $x = 0.74$ may be a root of $\cos x = x$ correct to two decimal places. Let us now check to see whether we have indeed located a root.

We again use $$g(x) = f(x) - x = \cos x - x,$$

and evaluate $g(0.74 \pm 0.005)$. Using a pocket calculator, we see that

$$g(0.735) > 0.006,$$

$$g(0.745) < -0.009,$$

and so the equation $g(x) = 0$ has a root $c$ such that $0.735 < c < 0.745$, by the intermediate value theorem (since $g$ is continuous on $[0.735, 0.745]$). Thus $c = 0.74$ correct to 2 decimal places, i.e. the equation $\cos x = x$ has a root $x = 0.74$ correct to two decimal places.

If the reader requires a root correct to six decimal places then it will be necessary to find the first forty terms of the sequence $(x_n)$. This will give a value of $c$ which can be checked by taking a very small interval about $c$ and applying the intermediate value theorem to $g$. For accuracy to six decimal places the intermediate value theorem has to be applied to the interval $[0.739\,084\,5, 0.739\,085\,5]$. Later, we will be able to prove (using the results of Chapter 5) that the equation $\cos x = x$ has just one real root.

## *EXERCISES 4.3.1*

**1** Show that the equation $$2 \sin x = x^2 - 1$$

has a solution between 1 and 2.

**2** Show that the equation $$2 \tan x = 1 + \cos x$$

has a solution in $(0, \pi/4)$.

**3** Show that the equation $$xe^x = 1$$

has a solution between 0 and 1. Write the equation as

$$x = e^{-x}.$$

Put $x_1 = 1$, $x_2 = e^{-x_1}$, ..., $x_{n+1} = e^{-x_n}$ $(n = 1, 2, \ldots)$. Use a calculator to find the first few terms of the sequence $(x_n)_{n=1}^{\infty}$. Hence, find a root of the equation correct to three decimal places. (Remember to use the intermediate value theorem to check that you really have found a root.) Use the same method to find a root correct to six decimal places.

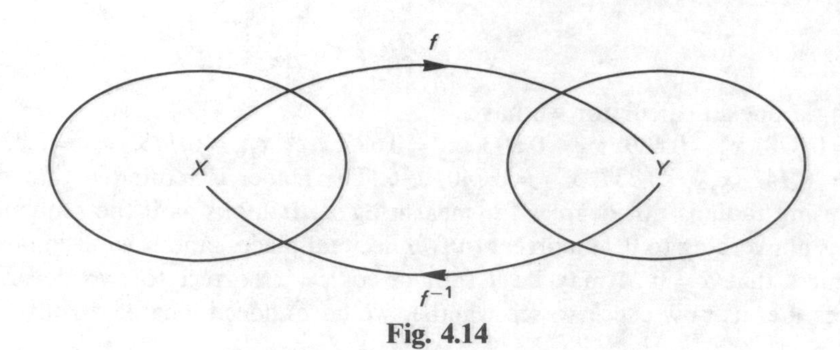


Fig. 4.14

Finally, we conclude with a section on the inverse function. Suppose $f: X \to Y$ is a function such that $Y = f(X)$ and $f(x_1) \neq f(x_2)$ whenever $x_1 \neq x_2$ and $x_1, x_2 \in X$, i.e. $f: X \to Y$ is a bijection. (For a more detailed explanation of the term bijection, the reader should consult the companion volume, *Guide to Abstract Algebra*.) Then for each $y \in Y$ there is one and only one $x \in X$ such that $y = f(x)$. The inverse function $f^{-1}$ from $Y$ onto $X$ is defined by $f^{-1}(y) = x$, where $x$ is the (unique) element of $X$ such that $y = f(x)$ (see Fig. 4.14). Thus for each $x \in X$,

$$f^{-1}(f(x)) = x$$

and for each $y \in Y$

$$f(f^{-1}(y)) = y.$$

We are interested in the question of whether continuity of $f$ implies continuity of $f^{-1}$. Since the existence of an inverse function requires that $f(x_1) \neq f(x_2)$ if $x_1 \neq x_2$ we will need to stipulate more than just continuity for $f$. It proves convenient to use strictly monotonic functions.

DEFINITION 4.3.1 Let $X, Y$ be two sets of real numbers. The function $f: X \to Y$ is said to be **strictly increasing** on $X$ if $f(x_1) < f(x_2)$ for all $x_1, x_2 \in X$ such that $x_1 < x_2$.

DEFINITION 4.3.2 Let $X, Y$ be sets of real numbers. The function $f: X \to Y$ is said to be **strictly decreasing** on $X$ if $f(x_1) > f(x_2)$ for all $x_1, x_2 \in X$ such that $x_1 < x_2$.

A **strictly monotonic** function is one which is either strictly increasing or strictly decreasing.

THEOREM 4.3.4 (Inverse Function Theorem) Suppose that $f(x)$ is defined for all $x$ such that $a \leqslant x \leqslant b$. Suppose also that $f$ is strictly increasing on

$[a,b]$, that $f$ is continuous on $[a,b]$, and that $f(a)=c$, $f(b)=d$. Then the inverse function $f^{-1}:[c,d]\to[a,b]$ exists, is strictly increasing and is continuous on $[c,d]$.

*Proof* Since $f$ is strictly increasing, it follows that $f(x_1)<f(x_2)$ for all $x_1, x_2$ such that $a\leqslant x_1<x_2\leqslant b$ and so $f(x_1)\neq f(x_2)$. Now $f(a)=c, f(b)=d$ and $f$ is continuous on $[a,b]$. By the intermediate value theorem every value between $c$ and $d$ is assumed on $[a,b]$, i.e., given any $y$ with $c\leqslant y\leqslant d$, there is some $x\in[a,b]$ such that

$$f(x)=y.$$

As $f$ is strictly increasing there is only one such value of $x$. The function $f^{-1}:[c,d]\to[a,b]$ is defined in the following way. For each $y$ such that $c\leqslant y\leqslant d$,

$$f^{-1}(y)=x, \tag{1}$$

where $x$ is the number such that $a\leqslant x\leqslant b$ and $f(x)=y$. Now we must prove that $f^{-1}$ is strictly increasing and continuous on $[c,d]$.

(I) $f^{-1}$ *is strictly increasing* If we draw a graph the result looks patently obvious (Fig. 4.15).

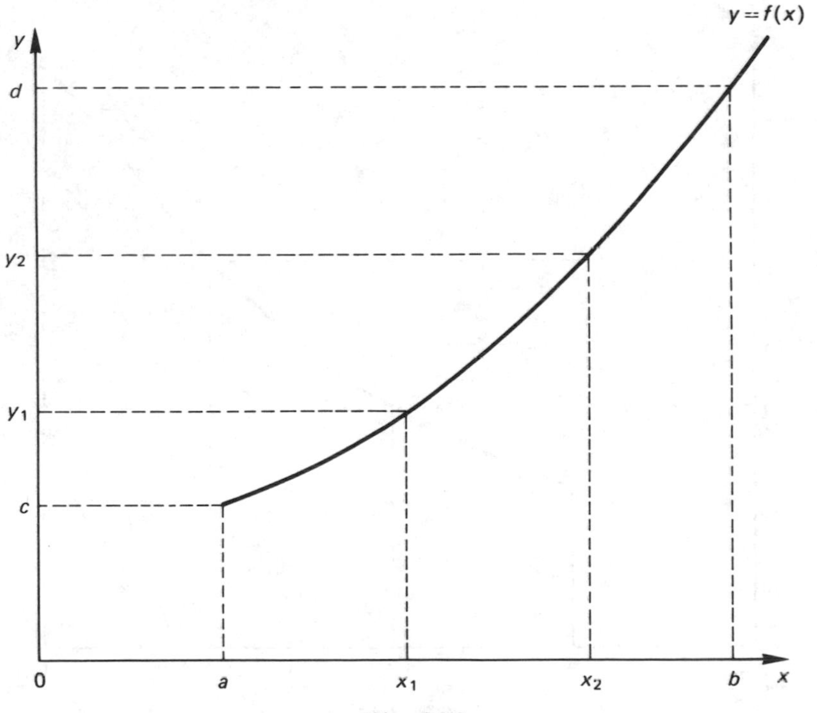


**Fig. 4.15**

Let $y_1, y_2$ be any two numbers such that $c \leqslant y_1 < y_2 \leqslant d$ and let $x_1 = f^{-1}(y_1)$, $x_2 = f^{-1}(y_2)$. Then, by definition, we have

$$y_1 = f(x_1),\ y_2 = f(x_2),$$

and we wish to prove that $x_1 < x_2$. There are only two possibilities; either (a) $a \leqslant x_2 \leqslant x_1 \leqslant b$ or (b) $a \leqslant x_1 < x_2 \leqslant b$. If condition (a) is satisfied, then $f(x_2) \leqslant f(x_1)$ because $f$ is strictly increasing, i.e. $y_2 \leqslant y_1$, which we know is untrue. Hence condition (a) is not satisfied and so $x_1$, $x_2$ must satisfy (b), i.e. $a \leqslant x_1 < x_2 \leqslant b$ and therefore $f^{-1}(y_1) < f^{-1}(y_2)$ whenever $c \leqslant y_1 < y_2 \leqslant d$. Hence $f^{-1}$ is strictly increasing.

(II) *$f^{-1}$ is continuous on $[c,d]$* First choose any number $y_1$ such that $c < y_1 < d$. To prove continuity at $y_1$ we have to show that given any $\varepsilon > 0$, $|f^{-1}(y) - f^{-1}(y_1)| < \varepsilon$ for all $y$ sufficiently close to $y_1$. If we write $x = f^{-1}(y)$, $x_1 = f^{-1}(y_1)$, then we need to prove that $|x - x_1| < \varepsilon$ whenever $y$ is sufficiently close to $y_1$. Figure 4.16 might prove helpful in disentangling the variables. What we have to prove is that $x$ is within $\varepsilon$ of $x_1$ if the corresponding $y$ is sufficiently close to $y_1$. The diagram suggests that sufficiently close to $y_1$ means within $\delta$ of $y_1$, where $\delta$ is the smaller of the numbers $h, k$. The reader may find it useful to keep referring to Fig. 4.16 during the execution of the proof in the next section.

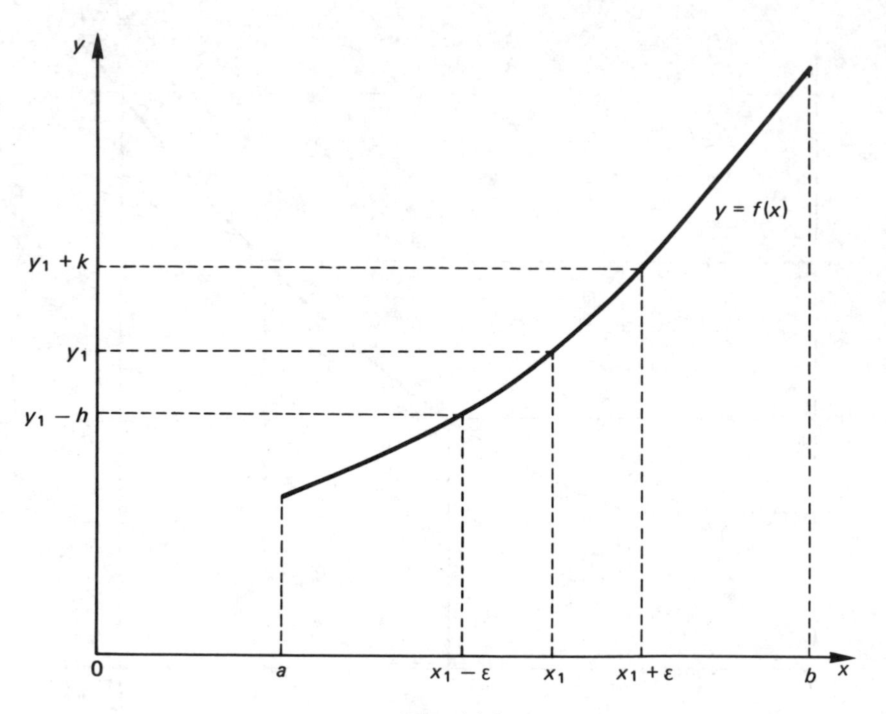


**Fig. 4.16**

To prove the result using symbols, let $x_1 = f^{-1}(y_1)$ and let $\varepsilon > 0$ be any positive real number such that $a \leqslant x_1 - \varepsilon < x_1 + \varepsilon \leqslant b$.

Let
$$f(x_1 - \varepsilon) = y_1 - h,$$
$$f(x_1 + \varepsilon) = y_1 + k.$$

Then $h > 0$ and $k > 0$ since $f$ is strictly increasing and we also have $c \leqslant y_1 - h < y_1 + k \leqslant d$.
Let $\delta$ be the smaller of the two numbers $h, k$. Then $\delta > 0$ and

$$c \leqslant y_1 - h \leqslant y_1 - \delta < y_1 + \delta \leqslant y_1 + k \leqslant d.$$

Now $f^{-1}$ is strictly increasing, and therefore if $y$ is any number such that

$$y_1 - \delta < y < y_1 + \delta$$

$$f^{-1}(y_1 - h) \leqslant f^{-1}(y_1 - \delta) < f^{-1}(y) < f^{-1}(y_1 + \delta) \leqslant f^{-1}(y_1 + k)$$

i.e. $$x_1 - \varepsilon \leqslant f^{-1}(y_1 - \delta) < f^{-1}(y) < f^{-1}(y_1 + \delta) \leqslant x_1 + \varepsilon.$$

Thus $|f^{-1}(y) - x_1| < \varepsilon$ for all $y$ such that $y_1 - \delta < y < y_1 + \delta$, i.e. $|f^{-1}(y) - f^{-1}(y_1)| < \varepsilon$ for all $y$ such that $|y - y_1| < \delta$. This means that if we are given any $\varepsilon > 0$ then there is a corresponding $\delta > 0$ such that $|f^{-1}(y) - f^{-1}(y_1)| < \varepsilon$ whenever $|y - y_1| < \delta$. Hence $f^{-1}$ is continuous at $y_1$, which is any point of $(c, d)$. A similar argument shows that $f^{-1}(y) \to f^{-1}(c)$ as $y \to c+$ and $f^{-1}(y) \to f^{-1}(d)$ as $y \to d-$. This completes the proof that $f^{-1}$ is continuous.

Naturally, this result remains true if the phrase 'strictly increasing' in the statement of the theorem is replaced by the phrase 'strictly decreasing' throughout.

* * * * *

This completes a long and somewhat difficult section on limits and continuity. The reader who has found the going tough may be pardoned for asking if there is an easier way to learn analysis. Such a question is not original. Ptolemy, who became ruler of Egypt after the death of Alexander the Great, found Euclid's elements tough reading and so he asked Euclid whether there was an easier way to a knowledge of geometry. Euclid made a very simple reply: 'There is no royal road to geometry.' Today we could make a similar pronouncement about analysis. There is no royal road to analysis. An understanding of analysis can be gained only with hard work, but it will prove worth the effort in the end. The final chapter of this volume furnishes some respite for the weary reader. It introduces the notion of differentiability and then illustrates some of its uses.

## *MISCELLANEOUS EXERCISES 4*

**1** The function $f: \mathbb{R} \to \mathbb{R}$ is given by

$$f(x) = \begin{cases} x^2+1 & (x<0), \\ [x] & (0 \leqslant x < 1), \\ 0 & (x=1), \\ x^2-1 & (x>1). \end{cases}$$

Determine whether the following limits exist:

(a) $\lim_{x \to 0+} f(x)$, (b) $\lim_{x \to 0} f(x)$, (c) $\lim_{x \to 1-} f(x)$,
(d) $\lim_{x \to 1} f(x)$.

When the limits exist find them. Decide where $f$ is continuous.

**2** The function $f: \mathbb{R} \to \mathbb{R}$ is given by

$$f(x) = \begin{cases} x & \text{if } x \text{ is rational,} \\ 0 & \text{if } x \text{ is irrational.} \end{cases}$$

Show that $f$ is continuous at the origin and discontinuous at all other points.

**3** The function $f: \mathbb{R} \to \mathbb{R}$ is given by $f(x) = x[x]$ for all real $x$, where $[x]$ is the greatest integer which is either less than $x$ or equal to $x$. Sketch a graph of $f$. Determine where $f$ is continuous.

**4** Let the functions $f, g$ both be continuous on $[a, b]$. Suppose that $f(a) < g(a)$ and $f(b) > g(b)$. Show that there is a number $c$ with $a < c < b$ such that

$$f(c) = g(c).$$

**5** Suppose the function $f$ has the property that $a \leqslant f(x) \leqslant b$ whenever $a \leqslant x \leqslant b$, i.e. $f: [a, b] \to [a, b]$. By considering the function $g$ given by $g(x) = f(x) - x$, show that there is a number $c$ such that $a \leqslant c \leqslant b$, and

$$f(c) = c.$$

Deduce that the equation $$x = \sin(e^x)$$

has a solution in $[-1, +1]$.

**6** The function $f: \mathbb{R} \to \mathbb{R}$ satisfies the relation $f(x+y) = f(x)f(y)$ for all $x, y \in \mathbb{R}$. Prove that

(a) $f(1) \geqslant 0$;
(b) if $f(1) = 0$, then $f(x) = 0$ for all $x$;
(c) if $f(1) = a > 0$, then $f(p) = a^p$ for all positive integers $p$, and $f(p/q) = a^{p/q}$ for all rationals $p/q$.

Show further that if $f(1) = a > 0$, and $f$ is continuous at each point of $\mathbb{R}$, then $f(x) = a^x$ for all $x \in \mathbb{R}$.

**7** Show that the equation

$$\sin^{19} x + \cos^{111} x = x$$

has a solution in the interval $[-2, 2]$.

# HINTS FOR SOLUTION OF EXERCISES

## Exercises 4.1.1

**3** Use Theorem 4.1.1.

## Exercises 4.1.4

**1** (a) and (b). Use algebra of limits. (c) Use Theorem 4.1.1.
(d) Use sequences. The following are suitable:

$$x_n = n\pi (n \in \mathbb{Z}^+), \qquad y_n = 2n\pi + \pi/2 \qquad (n \in \mathbb{Z}^+).$$

In this case $f(x_n) = 0$ and $f(y_n) = y_n^3/(y_n^2 + 1)$.
(e) and (f). Use same sequences as in (d).
(g) Use the inequality

$$|f(x)| \leqslant \frac{x}{x^2 + 1} \qquad (x > 0).$$

## Exercises 4.1.5

**1** Many of these parts use the algebra of limits. In addition the following points may be helpful:

(b) Remember, this is a limit as $x \to 1$.
(c) Remember that $\sin x/x \to 1$ as $x \to 0$.
(e) Use sequences. Suitable ones are $(x_n)$ and $(y_n)$, where

$$x_n = 1/n\pi \qquad \text{and} \qquad y_n = 1/(2n\pi + \pi/2)\ (n \in \mathbb{Z}^+).$$

In this case $f(x_n) = (-1)^n \sin(1/n\pi)$ and $f(y_n) = \cos[1/(2n\pi + \pi/2)]$.
(f) Use the identity $a^2 - b^2 = (a - b)(a + b)$ to deal with the difference of two square roots—see previous sets of exercises (Chapter 2).
(g) Deal with the cases $\alpha = 0$ and $\alpha \neq 0$ separately. Note that if $\alpha \neq 0$ then

$$\sin \alpha x/\alpha x \to 1 \qquad \text{as} \qquad x \to 0.$$

(i) Since $\sin x/x \to 1$ as $x \to 0$, there is some $\delta > 0$ such that $\frac{1}{2} < \sin x/x < \frac{3}{2}$ for $0 < |x| < \delta$. Hence if $0 < x < \delta$ then

$$\frac{\sin x}{x^2} > \frac{1}{2x}$$

and $f(x) \to \infty$ as $x \to 0+$. Similarly, $f(x) \to -\infty$ as $x \to 0-$.

**2** Find limits as $x \to 0+$ and as $x \to 0-$ from the definition of $f$.

### Exercises 4.2.1

**2** (a) Use the fact that $f(x) = -1$ for $-1 \leqslant x < 0$.
(c) Use $|f(x)| \leqslant \sqrt{|x|}$ for $x \neq 0$.

**3 and 4** Use sequences.

**5** (a) Use sequences.

### Miscellaneous Exercises 4

**2** Use sequences.

**3** If $n \in \mathbb{Z}$ and $n \leqslant x < n+1$ then $[x] = n$ and $f(x) = nx$ for $n \leqslant x < n+1$.

**6** (a) Use $f(1) = f(\frac{1}{2} + \frac{1}{2}) = [f(\frac{1}{2})]^2$.
(b) Use $f(x) = f(x-1+1) = f(x-1)f(1)$.

Finally, to prove that $f(x) = a^x$ for all real $x$, use a sequence of rational numbers which converges to $x$.

## ANSWERS TO EXERCISES

### Exercises 4.1.4

**1** (a) $f(x) \to \frac{2}{3}$. (b) $f(x) \to -1$. (c) $f(x) \to \infty$.
(d) No limit. (e) No limit. (f) No limit.
(g) $f(x) \to 0$. (h) $f(x) \to \infty$. (i) $f(x) \to 0$.

**2** The following are just some possible examples:

(a) $f(x) = x$ and $g(x) = \sin x$. Use sequences to show that $f(x)g(x)$ does not tend to any limit as $x \to \infty$. The sequences $(x_n)$ and $(y_n)$ would be suitable, where $x_n = n\pi$ and $y_n = 2n\pi + \pi/2$.
(b) If $f(x) = x^2$ and $g(x) = 1/(1+x^2)$, then $f(x)g(x) \to 1$ as $x \to \infty$. If $f(x) = x^3$ and $g(x) = 1/(1+x^2)$, then $f(x)g(x) \to \infty$ as $x \to \infty$.

**4** The following are some possible examples:

(a) $f(x) = x$ and $g(x) = -x + \sin x$. (b) $f(x) = x$ and $g(x) = -x - x^2$.
(c) $f(x) = x + l$ and $g(x) = -x$.

### Exercises 4.1.5

**1** (a) $f(x) \to 1$. (b) $f(x) \to \frac{3}{7}$. (c) $f(x) \to 1$.
(d) $f(x) \to 0$. (e) No limit. (f) $f(x) \to 1$.
(g) $f(x) \to \alpha$. (h) No limit. (i) $f(x) \to 0$.
(j) No limit.

**2** As $x \to 0$, $f(x) \to 0$. As $x \to 1$, $f(x) \to 1$.

**3** The following are some suitable examples:

(a) $f(x) = x - 1$ and $g(x) = (x-1)\sin(x-1)$.
(b) $f(x) = x - 1$ and $g(x) = (x-1)^3$.
(c) $f(x) = 3(x-1)$ and $g(x) = (x-1)$.

## Exercises 4.2.1

**2** (a) $f$ is not continuous at the origin. (b) $f$ is continuous at the origin.
(c) $f$ is continuous at the origin.

**5** (b) One possible example is the function $f$ given by

$$f(x) = \begin{cases} \sqrt{2} & \text{when } x \text{ is rational,} \\ x & \text{when } x \text{ is irrational.} \end{cases}$$

## Miscellaneous Exercises 4

**1** (a) $f(x) \to 0$ as $x \to 0+$. (b) No limit.
(c) $f(x) \to 0$ as $x \to 1-$. (d) $f(x) \to 0$ as $x \to 1$.

The function $f$ is discontinuous at the origin and continuous at all other points of $\mathbb{R}$.

**3** $f$ is continuous at all points of $\mathbb{R}\backslash\mathbb{Z}$ and also continuous at the origin. It is discontinuous at all other points.

**4** Apply the intermediate value theorem to $f - g$ on the interval $[a, b]$.

# 5 DIFFERENTIABLE FUNCTIONS

## Differentiable functions, Rolle's theorem, mean value theorem, Taylor's theorem, power series expansions of functions, l'Hôpital's rule, maxima and minima

### 5.1 DIFFERENTIATION

Calculus originated with the work of two celebrated mathematicians, both born in the seventeenth century. One of these was Sir Isaac Newton (1642–1727), the other was Baron Gottfried Wilhelm von Leibniz (1646–1716).

Newton's extraordinary ability was noticed early in his life by a maternal uncle who encouraged his mother to enter him for Cambridge. As a result, Newton entered Trinity College, Cambridge, in 1661. Soon after he gained his degree the college was temporarily closed because of the plague and Newton returned home to continue his work. This period of enforced absence from Cambridge in 1665–6 saw the birth of many new ideas which laid the foundation for Newton's subsequent work. In particular, the period was marked by the beginnings of calculus and the emergence of infinite series as a tool in mathematics. During the next few years Newton wrote several accounts of his work. The first of these was the treatise already mentioned in earlier chapters, *De analysi per aequationes numero terminorum infinitas*, which he wrote in 1669. Initially the manuscript was used mainly for circulation among his friends and its publication was delayed until 1711. In fact, Newton's first published account of calculus did not appear until 1687. It was his celebrated *Philosophiae naturalis principia mathematica*—frequently referred to in literature as 'Newton's *Principia*'. Newton's actual terminology might seem a little strange to us today and his notation, though not totally unfamiliar, is not the one in most common use. However, this work does include some of the familiar rules for differentiating products, quotients, etc. which we will derive later in the chapter, though we will not use Newton's notation.

Leibniz was born in Leipzig and entered the local university at the early age of fifteen, to study theology, law, philosophy and mathematics. After obtaining his doctorate at the University of Altdorf, he was offered a professorship in law, which he declined in favour of a career in the diplomatic service. During his travels abroad he maintained an interest in mathematics and became a member of the Royal Society when he visited London in 1673. At that time, manuscripts of Newton's *De Analysi* were in existence and it would have been possible for him to have obtained one. It is, however, very doubtful whether he could have gained much from reading such a manuscript as he lacked much of the necessary background knowledge of mathematics at that stage.

The year 1684 was marked by the appearance of the first published accounts of differential calculus written by Leibniz. It is known that he developed his theory of differential calculus between his visits to London in 1673 and 1676, and it is generally accepted nowadays that Newton and Leibniz worked completely independently. In the first edition of his *Principia* (1687) Newton acknowledged Leibniz's contribution to calculus. The following years saw heated arguments between adherents of Newton and Leibniz about whether Leibniz had been guilty of plagiarism and the Royal Society was asked to adjudicate. When the third edition of Newton's *Principia* appeared (1726) he had deleted all reference to Leibniz. The most serious consequence of these disagreements was the rift it produced between mathematicians in England and those on the Continent. At a time when new disciplines like analysis were beginning to emerge on the Continent, English mathematicians were virtually isolated—a situation which was not reversed until the early nineteenth century.

Leibniz was a firm believer in the vital importance of clear, comprehensible notation. He was the one to introduce the 'd' notation and the calculus of today owes its form primarily to the influence of Leibniz. Like Newton he derived formulae for derivatives of products and quotients. He is particularly remembered by generations of students for his formula for the $n$th-order derivative of the product of two functions—a formula which is called Leibniz's rule to this day.

In the sixth form differentiation is normally introduced in the following way. Write $y=f(x)$ and $y+\delta y=f(x+\delta x)$, where $\delta x\neq 0$. Then subtract to obtain

$$\delta y=f(x+\delta x)-f(x),$$

which gives

$$\frac{\delta y}{\delta x}=\frac{f(x+\delta x)-f(x)}{\delta x}.$$

If we can draw the graph $y=f(x)$, then the ratio $\delta y/\delta x$ measures the gradient of the chord $PQ$ (see Fig. 5.1), where $P$ is the point with coordinates $(x, y)$ and $Q$ has coordinates $(x+\delta x, y+\delta y)$. Thus $(\delta y/\delta x)=\tan\alpha$, where $\alpha$ is the angle which the chord $PQ$ makes with the positive direction of the $x$-axis. The final step is to consider what happens to $(\delta y/\delta x)$ as $\delta x\to 0$, i.e. we need to consider whether $[f(x+\delta x)-f(x)]/\delta x$ tends to a limit as $\delta x\to 0$. This

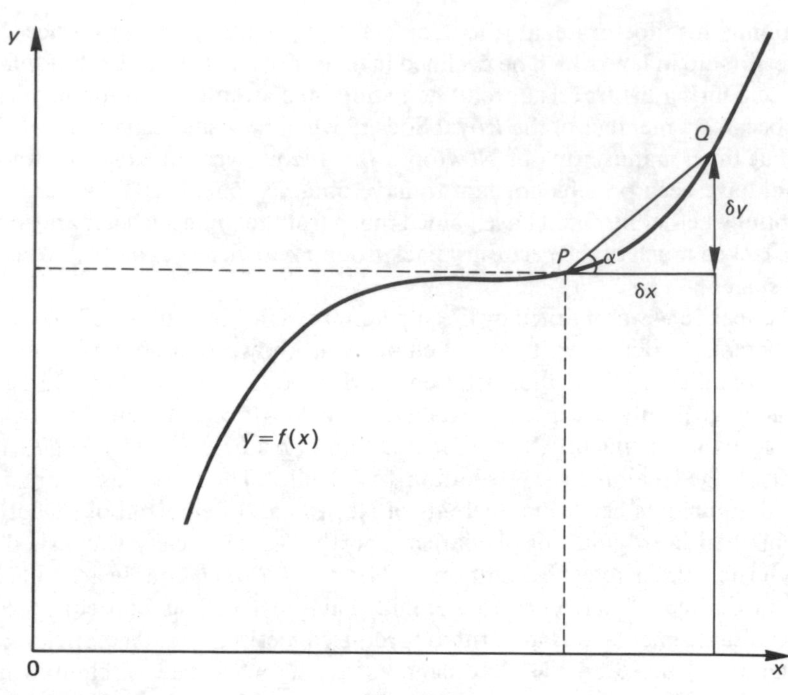


Fig. 5.1

should not present an insurmountable problem as we know how to deal with limits of functions.

The whole process we have outlined can be carried out using only symbols without any recourse to a diagram at all. Of course, if it is to make sense, then $f(x)$ must be defined and $f(x+\delta x)$ must be defined for all sufficiently small $\delta x$. This means we will have to stipulate that the domain of definition of $f$ has certain properties, but this will not cause undue difficulty. Normally, in this context, the symbol $h$ is used in place of $\delta x$, and we therefore consider what happens to

$$\frac{f(x+h)-f(x)}{h}$$

as $h \to 0$. Finally, in order to ensure that our definition is precise, we restrict our attention to the behaviour of $f$ near a particular point $x_1$.

DEFINITION 5.1.1 Suppose $f(x)$ is defined for all $x$ such that $x_1 - R < x < x_1 + R$, where $R$ is some given positive real number, i.e. $f$ is defined for all points sufficiently close to $x_1$. If the quotient

$$\frac{f(x_1+h)-f(x_1)}{h}$$

tends to a finite limit as $h \to 0$, then $f$ is said to be **differentiable at** $x_1$. The value of this finite limit is then called the **derivative of** $f$ **at** $x_1$ and it is denoted by $f'(x_1)$,

i.e.
$$f'(x_1) = \lim_{h \to 0} \frac{f(x_1 + h) - f(x_1)}{h}.$$

DEFINITION 5.1.2 Suppose $f(x)$ is defined for all $x$ such that $a < x < b$. Then $f$ is said to be **differentiable on the open interval** $(a, b)$, if $f$ is differentiable at all points of the interval.

The reader will notice that Definition 5.1.1 includes an explicit statement concerning the domain of definition of $f$. It is assumed that this domain contains some interval $(x_1 - R,\ x_1 + R)$ $(R > 0)$. Whenever differentiability is used it is taken for granted that some such condition is satisfied unless it is explicitly stated otherwise.

Sometimes it is convenient to use $x$ in place of $x_1 + h$. In this case $h$ is replaced by $x - x_1$, and the condition $h \to 0$ is replaced by the condition $x \to x_1$. With this notation,

$$f'(x_1) = \lim_{x \to x_1} \frac{f(x) - f(x_1)}{x - x_1}.$$

Indeed, the definition of differentiability is frequently stated using just this notation. The statement would then read as follows.

If the quotient
$$\frac{f(x) - f(x_1)}{x - x_1}$$
tends to a finite limit as $x \to x_1$, then $f$ is said to be differentiable at $x_1$.

Now suppose that $f$ is differentiable at $x_1$. Then

$$\lim_{x \to x_1} \frac{f(x) - f(x_1)}{x - x_1} = f'(x_1)$$

and hence

$$\lim_{x \to x_1} [f(x) - f(x_1)] = \lim_{x \to x_1} \left[\frac{f(x) - f(x_1)}{x - x_1}\right] \lim_{x \to x_1} (x - x_1) = 0$$

using the algebra of limits.

i.e.
$$\lim_{x \to x_1} f(x) = f(x_1).$$

Thus $f$ is continuous at $x_1$ and we have shown that differentiability implies continuity. This result is now stated in the form of a theorem.

THEOREM 5.1.1 Suppose that $f(x)$ is defined for all $x$ such that

$x_1 - R < x < x_1 + R$, where $R$ is some positive real number. If $f$ is differentiable at $x_1$, then $f$ is continuous at $x_1$.

COROLLARY Suppose $f(x)$ is defined for all $x$ such that $a < x < b$. If $f$ is differentiable on $(a, b)$, then $f$ is continuous on $(a, b)$.

There are, however, continuous functions which are not differentiable and the converse of Theorem 5.1.1 is not true. For example, let $f: \mathbb{R} \to \mathbb{R}$ be defined by $f(x) = |x|$. Then $f$ is clearly continuous at the origin since $f(x) \to 0 = f(0)$ as $x \to 0$. Now if $x > 0$,

$$\frac{f(x) - f(0)}{x - 0} = \frac{x - 0}{x - 0} = 1$$

and
$$\lim_{x \to 0+} \frac{f(x) - f(0)}{x - 0} = 1. \tag{1}$$

However, if $x < 0, \dfrac{f(x) - f(0)}{x - 0} = \dfrac{|x| - |0|}{x - 0} = \dfrac{-x - 0}{x - 0} = -1$

and
$$\lim_{x \to 0-} \frac{f(x) - f(0)}{x - 0} = -1. \tag{2}$$

From equations (1) and (2) we see that

$$\frac{f(x) - f(0)}{x - 0}$$

does not tend to any limit as $x \to 0$. Thus $f$ is not differentiable at the origin. The graph of this particular function is easily sketched (see Fig. 5.2). It has

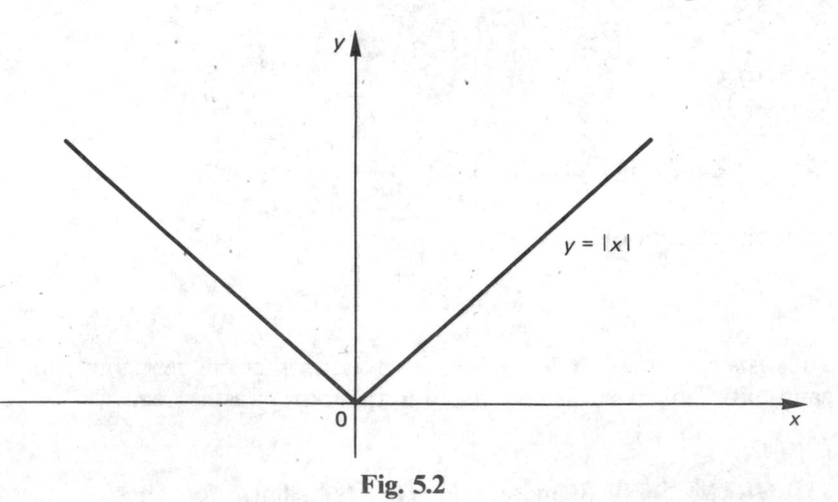


Fig. 5.2

a V-shape at the origin. The slope is $+1$ on the right of the origin and $-1$ on the left, which are the right and left limits we obtained in equations (1) and (2).

This is just one example used to illustrate the fact that continuity does not imply differentiability. It is a particularly simple example, in which the derivative does not exist at the origin, but it exists at all other points. There are far more complicated situations than this one. There are, for example, functions $f: \mathbb{R} \to \mathbb{R}$ which are continuous at all points of $\mathbb{R}$ and not differentiable anywhere at all. Such functions are continuous everywhere, but their graphs cannot be sketched as they do not have a tangent at any point at all. It is therefore fortunate that our definitions of concepts like continuity are not dependent on geometric representations.

In their first accounts of the theory of differential calculus, Newton and Leibniz both included rules for differentiating sums, products, etc. These familiar rules can now be proved without any difficulty.

THEOREM 5.1.2 Suppose that $f(x)$ and $g(x)$ are both defined for $x_1 - R < x < x_1 + R$ where $R$ is some given positive real number. If $f, g$ are differentiable at $x_1$, then

(a) $f+g$ is differentiable at $x_1$, with derivative $f'(x_1)+g'(x_1)$;
(b) $fg$ is differentiable at $x_1$, with derivative $f(x_1)g'(x_1)+f'(x_1)g(x_1)$;
(c) $f/g$ is differentiable at $x_1$ with derivative

$$\frac{f'(x_1)g(x_1)-f(x_1)g'(x_1)}{[g(x_1)]^2},$$

provided $g(x_1) \neq 0$.

*Proof* Since $f, g$ are differentiable at $x_1$,

$$\frac{f(x)-f(x_1)}{x-x_1} \to f'(x_1), \tag{1}$$

$$\frac{g(x)-g(x_1)}{x-x_1} \to g'(x_1) \tag{2}$$

as $x \to x_1$. Moreover, $f, g$ are both continuous at $x_1$, since differentiability implies continuity,

i.e.
$$f(x) \to f(x_1) \tag{3}$$

$$g(x) \to g(x_1) \tag{4}$$

as $x \to x_1$.

(a) Let $0 < |x - x_1| < R$. Then

$$\frac{[f(x)+g(x)]-[f(x_1)+g(x_1)]}{x-x_1} = \frac{f(x)-f(x_1)}{x-x_1} + \frac{g(x)-g(x_1)}{x-x_1}$$

and this tends to $f'(x_1)+g'(x_1)$ as $x \to x_1$, using the algebra of limits and relations (1) and (2). Hence $f+g$ is differentiable at $x_1$ and its derivative at $x_1$ is $f'(x_1)+g'(x_1)$.

(b) Let $0<|x-x_1|<R$. Then

$$\frac{f(x)g(x)-f(x_1)g(x_1)}{x-x_1}=\frac{f(x)g(x)-f(x_1)g(x)+f(x_1)g(x)-f(x_1)g(x_1)}{x-x_1}$$

$$=g(x)\frac{f(x)-f(x_1)}{x-x_1}+f(x_1)\frac{g(x)-g(x_1)}{x-x_1}$$

and this tends to $g(x_1)f'(x_1)+f(x_1)g'(x_1)$ as $x \to x_1$ using the algebra of limits and (1), (2) and (4). Thus the product $fg$ is differentiable at $x_1$ and its derivative at $x_1$ is

$$f'(x_1)g(x_1)+f(x_1)g'(x_1).$$

(c) Since $g$ is continuous at $x_1$ and $g(x_1) \neq 0$, there is some $\delta>0$ (with $\delta \leqslant R$) such that $g(x) \neq 0$ for all $x$ such that $x_1-\delta<x<x_1+\delta$, by Theorem 4.2.1. Let $0<|x-x_1|<\delta$. Then

$$\frac{\dfrac{1}{g(x)}-\dfrac{1}{g(x_1)}}{x-x_1}=\frac{g(x_1)-g(x)}{g(x)g(x_1)(x-x_1)}$$

$$=-\frac{1}{g(x)}\frac{1}{g(x_1)}\frac{g(x)-g(x_1)}{x-x_1}$$

and this tends to $-g'(x_1)/[g(x_1)]^2$ as $x \to x_1$, using the algebra of limits and relations (2) and (4). Thus $1/g$ is differentiable at $x_1$, and its derivative at this point is

$$\frac{-g'(x_1)}{[g(x_1)]^2}.$$

Using the rule for differentiating a product, we see that $f/g$ is differentiable at $x_1$; its derivative is

$$\frac{f'(x_1)g(x_1)-f(x_1)g'(x_1)}{[g(x_1)]^2}.$$

This completes the proof.

Up to this point we have taken great care to state explicitly in each theorem that the functions under consideration are defined on certain specified intervals. Since continuity and differentiability are defined on the understanding that the domain of definition of the functions satisfies certain conditions, such care is not really necessary. In future we will omit such details, and will take it for granted that the concepts of continuity and differentiability are now sufficiently familiar to the reader that such reminders are no longer needed.

We can now differentiate sums, products and quotients of differentiable functions. Can we manage to differentiate more complicated functions like $\cos(e^x)$? If we are to handle such functions then we must have a rule for differentiating composite functions.

THEOREM 5.1.3 Suppose that the function $g$ is differentiable at $x_1$ and $g(x_1) = y_1$. Suppose also that the function $f$ is differentiable at $y_1$. Then the composite function $f \circ g$ is differentiable at $x_1$ and its derivative at $x_1$ is

$$f'(y_1)g'(x_1).$$

(If we write $y = g(x)$, $u = f(y)$ so that $u = f(g(x))$, then the above rule can be written as

$$\frac{du}{dx} = \frac{du}{dy} \cdot \frac{dy}{dx}$$

since $\dfrac{du}{dy} = f'(y)$ and $\dfrac{dy}{dx} = g'(x)$, which is the familiar rule for differentiating a function of a function.)

*Proof* Since $g$ is differentiable at $x_1, g(x)$ is defined for all $x$ sufficiently close to $x_1$ and $g$ is continuous at $x_1$. Using the fact that $f$ is differentiable at $y_1$, we can conclude that $f(g(x))$ is defined for all $x$ such that $x_1 - R < x < x_1 + R$ (where $R$ is some given positive real number)—see the section preceding Theorem 4.2.4.

Since $f$ is differentiable at $y_1$,

$$\frac{f(y_1 + k) - f(y_1)}{k} \to f'(y_1)$$

as $k \to 0$. We define the function $\varepsilon$ by $\varepsilon(k) = 0$ when $k = 0$ and

$$\varepsilon(k) = \frac{f(y_1 + k) - f(y_1)}{k} - f'(y_1) \qquad (k \neq 0). \tag{1}$$

Then
$$f(y_1 + k) - f(y_1) = kf'(y_1) + k\varepsilon(k), \tag{2}$$

where $\varepsilon(k) \to 0$ as $k \to 0$.

Now let $0 < |h| < R$, and write

$$g(x_1 + h) = y_1 + k$$

so that
$$k = g(x_1 + h) - g(x_1), \tag{3}$$

because $y_1 = g(x_1)$. Then we have

$$f(g(x_1+h)) - f(g(x_1)) = f(y_1+k) - f(y_1)$$
$$= kf'(y_1) + k\varepsilon(k)$$

by (2). Hence
$$\frac{f(g(x_1+h)) - f(g(x_1))}{h} = \frac{k}{h}f'(y_1) + \frac{k}{h}\varepsilon(k). \tag{4}$$

Now, by definition,
$$\frac{k}{h} = \frac{g(x_1+h) - g(x_1)}{h},$$

and therefore
$$\frac{k}{h} \to g'(x_1) \tag{5}$$

as $h \to 0$. Moreover $g$ is continuous at $x_1$ and so $k = g(x_1+h) - g(x_1) \to 0$ as $h \to 0$. Thus as $h \to 0$, we must have $k \to 0$ which in turn implies that $\varepsilon(k) \to 0$. From (4) it now follows that as $h \to 0$

$$\frac{f(g(x_1+h)) - f(g(x_1))}{h} \to g'(x_1)f'(y_1)$$

and so $f \circ g$ is differentiable at $x_1$ with derivative $f'(y_1)g'(x_1)$. This completes the proof.

The above proof was rather long and involved. The reader who is wide awake might be tempted to ask whether we could use a more straightforward method. Using the same notation it is rather tempting to use the following argument. Write

$$\frac{f(g(x_1+h)) - f(g(x_1))}{h} = \frac{f(y_1+k) - f(y_1)}{k} \cdot \frac{k}{h}. \tag{*}$$

Now $k/h \to g'(x_1)$ as $h \to 0$ and it all looks plain sailing. Unfortunately there is one serious snag. We do not know whether there are arbitrarily small non-zero $|h|$ for which the corresponding value of $k$ is zero. The relation (*) does not make sense if $k = 0$. To allow for the possibility of such an occurrence, we have been forced to use the longer argument.

Finally we need a rule for differentiating inverse functions.

THEOREM 5.1.4 Suppose the function $f$ is continuous and strictly increasing on the closed interval $[x_1 - R, x_1 + R]$ where $R$ is some given positive real number. Suppose also that $f$ is differentiable at $x_1$ and $f'(x_1) \neq 0$. Let $y_1 = f(x_1)$. Then the inverse function is differentiable at $y_1$ and its derivative at that point is $1/f'(x_1)$.

*Proof* Let $c = f(x_1 - R), d = f(x_1 + R)$, then $f$ has an inverse function which is continuous and strictly increasing on $[c, d]$ by Theorem 4.3.4. We will use

$g$ for this inverse function rather than $f^{-1}$ to prevent confusion with superscripts.

Let $k$ be any non-zero number such that $y_1 + k$ lies in $[c, d]$, and let

$$g(y_1 + k) = x_1 + h. \tag{1}$$

Then $h \neq 0$ since $g$ is strictly increasing and

$$y_1 + k = f(x_1 + h). \tag{2}$$

Moreover

$$\frac{g(y_1 + k) - g(y_1)}{k} = \frac{(x_1 + h) - x_1}{k} = \frac{h}{k}$$

$$= \frac{h}{f(x_1 + h) - f(x_1)}.$$

Now $k \to 0$ implies $h \to 0$ as $g$ is continuous and the condition $h \to 0$ in turn implies that

$$\frac{f(x_1 + h) - f(x_1)}{h} \to f'(x_1).$$

Thus as $k \to 0$

$$\frac{g(y_1 + k) - g(y_1)}{k} \to \frac{1}{f'(x_1)}$$

and the inverse function $g$ is differentiable at $y_1$, with derivative $1/f'(x_1)$.

This completes the proof.

We have shown that $g'(y_1) = 1/f'(x_1)$. Now, using the notation $y = f(x)$ and $x = g(y)$, we see that

$$f'(x) = \frac{dy}{dx}, \qquad g'(y) = \frac{dx}{dy}.$$

The relation $g'(y_1) = 1/f'(x_1)$, therefore, gives the familiar rule

$$\frac{dx}{dy} = 1 \Big/ \frac{dy}{dx}.$$

The result in Theorem 5.1.4 remains true if the phrase 'strictly increasing' is replaced by the phrase 'strictly decreasing' throughout the statement of the theorem.

---

*Example 5.1.1*

The functions $f: \mathbb{R} \to \mathbb{R}$ and $g: \mathbb{R} \to \mathbb{R}$ are defined by

$$f(x) = \begin{cases} x \sin(1/x) & (x \neq 0), \\ 0 & (x = 0); \end{cases} \qquad g(x) = \begin{cases} x^2 \sin(1/x) & (x \neq 0), \\ 0 & (x = 0). \end{cases}$$

Decide where $f, g$ are (a) continuous, (b) differentiable. Find the derivatives where they exist.

*Solution*

(a) In Example 3 at the end of section 4.2 we showed that $f$ is continuous at every point of $\mathbb{R}$. Moreover, $g(x) = xf(x)$ for all $x \in \mathbb{R}$ and hence $g$ is also continuous at all points of $\mathbb{R}$, since the product of two continuous functions is continuous.

(b) Since $1/x$ is differentiable at all points of $\mathbb{R}\backslash\{0\}$ and the sine is differentiable at all points of $\mathbb{R}$, it follows from the law for differentiating composite functions that the derivative of $\sin(1/x)$ is

$$-\frac{1}{x^2}\cos\left(\frac{1}{x}\right) \qquad (x \neq 0).$$

Using the rule for products we see that

$$f'(x) = \sin(1/x) - \frac{1}{x}\cos\left(\frac{1}{x}\right) \qquad (x \neq 0)$$

$$g'(x) = 2x\sin(1/x) - \cos(1/x) \qquad (x \neq 0).$$

Thus $f, g$ are both differentiable at all points of $\mathbb{R}\backslash\{0\}$.

The question of behaviour at the origin still remains. If $x \neq 0$

$$\frac{f(x) - f(0)}{x - 0} = \frac{x\sin(1/x)}{x} = \sin(1/x)$$

and we know that $\sin(1/x)$ does not tend to any limit as $x \to 0$. Hence $f$ is not differentiable at the origin.

If $x \neq 0$, $$\frac{g(x) - g(0)}{x - 0} = \frac{x^2\sin(1/x)}{x} = x\sin(1/x).$$

and $x\sin(1/x) \to 0$ as $x \to 0$. Thus $g$ is also differentiable at the origin with $g'(0) = 0$.

---

## EXERCISES 5.1.1

**1** Use the identity

$$\cos x - \cos a = 2\sin\frac{x+a}{2}\sin\frac{a-x}{2}$$

to show that $\cos x$ is differentiable at every point $x \in \mathbb{R}$ and its derivative is $-\sin x$. (You may use the fact that $(\sin t)/t \to 1$ as $t \to 0$.)

**2** The function $f\colon \mathbb{R} \to \mathbb{R}$ is defined by

$$f(x) = \begin{cases} e^x & (x < 0), \\ x + 1 & (x \geqslant 0). \end{cases}$$

Show that $f$ is differentiable at all points of $\mathbb{R}$.

**3** Find the derivative of each of the following functions $f$. In each case state explicitly the values of $x$ for which the formula for $f'$ is valid.

(a) $f(x)=x^3 \sinh x \sin(x^2)$; (b) $f(x)=e^{x\sin x}$;
(c) $f(x)=\cos[\log(1+x^2)]$; (d) $f(x)=\sinh^{-1}(1+x^2)$;
(e) $f(x)=x^{\log x}$ $(x>0)$.

## Higher-order derivatives

Suppose the function $f$ is differentiable at all points of the open interval $(a-R,\ a+R)$, where $R$ is some positive real number. Then $f'(x)$ is defined for all $x$ such that $a-R<x<a+R$ and we could ask the question: Is $f'$ differentiable at $a$? If $[f'(x)-f'(a)]/(x-a)$ tends to a finite limit as $x \to a$, then $f'$ is itself differentiable at $a$. Its derivative will be denoted by $f''(a)$, which is normally called the second-order derivative of $f$ at $a$. Higher-order derivatives can be defined in the same way. There are no new principles involved.

DEFINITION 5.1.3 Suppose the function $f$ is differentiable at all points of an open interval $(a-R,\ a+R)$, where $R$ is some positive real number. If the derivative $f'$ is itself differentiable at $a$, then $f$ is said to possess **a second-order derivative at** $a$, which is denoted by $f''(a)$.

Thus
$$f''(a)=\lim_{x\to a}\frac{f'(x)-f'(a)}{x-a}.$$

This can be generalised to give the definition of an $n$th-order derivative.

DEFINITION 5.1.4 Suppose the function $f$ possesses an $(n-1)$th-order derivative at all points of an open interval $(a-R,\ a+R)$, where $R$ is some positive real number and $n$ is positive integer with $n \geqslant 2$. If the $(n-1)$th-order derivative $f^{(n-1)}$ is differentiable at $a$, then its derivative is called the $n$th-**order derivative of** $f$ **at** $a$ and denoted by $f^{(n)}(a)$. Thus
$$f^{(n)}(a)=\lim_{x\to a}\frac{f^{(n-1)}(x)-f^{(n-1)}(a)}{x-a}.$$

DEFINITION 5.1.5 If $f$ possesses an $n$th-order derivative at all points of an open interval $(a,b)$, we say that $f$ is $n$ **times differentiable** on $(a,b)$.

As one might expect, slight modifications are made when dealing with a closed interval. In the case of continuity of a function $f$ on a closed interval $[a,b]$ we used only one-sided limits at the end-points $a,b$ in order to avoid considering what happens to $f(x)$ when $x$ is outside $[a,b]$. The same convention holds for differentiability on a closed interval.

DEFINITION 5.1.6 Let $f(x)$ be defined for all $x$ such that $a \leqslant x \leqslant b$. If $f$ is differentiable at all points of the open interval $(a,b)$ and the one-sided limits

$$\lim_{x\to a+} \frac{f(x)-f(a)}{x-a}, \qquad \lim_{x\to b-} \frac{f(x)-f(b)}{x-b}$$

both exist and are finite, then $f$ is said to be **differentiable on the closed interval** $[a,b]$. In these circumstances, it is normal to write $f'(a)$ for

$$\lim_{x\to a+} \frac{f(x)-f(a)}{x-a}$$

and $f'(b)$ for

$$\lim_{x\to b-} \frac{f(x)-f(b)}{x-a}.$$

Naturally, if the domain of definition of $f$ extends outside $[a,b]$ and if the normal derivatives exist at $a$ and $b$, they have the same value as the respective one-sided limits and the end-points need no special treatment. For example, we have seen that if $f(x)=\cos x$, then $f$ is differentiable at all points of $\mathbb{R}$, and therefore it is differentiable at all points of any closed interval $[a,b]$. Hence $f$ is differentiable on $[a,b]$. In this case there was no special problem at the end-points.

The same convention is adopted for higher-order derivatives on a closed interval. If $n$ is a positive integer with $n \geqslant 2$, then $f$ is said to possess an $n$th-order derivative on a closed interval $[a,b]$ if the $(n-1)$th-order derivative $f^{(n-1)}$ exists on $[a,b]$ and is differentiable on $[a,b]$.

We are now the proud possessors of a rigorous theory of differentiation, but have we any differentiable functions on which to try it out? Perhaps the most obvious place in which to commence our search is the stock of continuous functions from Chapter 4. Clearly the function $f\colon \mathbb{R}\to\mathbb{R}$ given by $f(x)=x$ is differentiable at all points of $\mathbb{R}$. For if $x_0\in\mathbb{R}$ and $x\neq x_0$, then

$$\frac{f(x)-f(x_0)}{x-x_0}=\frac{x-x_0}{x-x_0}=1$$

and

$$\frac{f(x)-f(x_0)}{x-x_0}\to 1$$

as $x\to x_0$. Hence $f$ is differentiable at all points of $\mathbb{R}$ and $f'(x_0)=1$ for all $x_0\in\mathbb{R}$. Similarly, we can easily show that a constant function is differentiable with derivative zero. Since sums and products of differentiable functions are differentiable, it follows that positive integral powers are differentiable everywhere, and the derivative of $x^n$ is $nx^{n-1}$. Thus polynomials are differentiable at all points of $\mathbb{R}$. Hence a rational function (i.e. the quotient of two polynomials) is differentiable at all points at which the denominator is not zero. Using the fact that $\sin t/t \to 1$ as $t \to 0$, we can prove that the sine and cosine are differentiable everywhere and have the familiar derivatives.

## 5.2 ROLLE'S THEOREM AND THE MEAN VALUE THEOREM

In the previous section we showed how the technical rules for differentiation can be derived using as a basis the theory of limits. This section contains some rather pretty results for differentiable functions.

The first result is named after the French mathematician Michel Rolle (1652–1719). It is ironic that Rolle should above all be remembered in connection with a theorem concerning differentiation, as he was never impressed with calculus. In fact, he is supposed to have dismissed the subject as 'an ingenious collection of fallacies'. His result was originally published in 1691 in an obscure book on geometry and algebra. It is really very simple, but its consequences are far-reaching.

THEOREM 5.2.1 (Rolle's Theorem) Suppose that

(i) the function $f$ is continuous on the closed interval $[a, b]$,
(ii) $f$ is differentiable on the open interval $(a, b)$,
(iii) $f(a) = f(b)$.

Then there is some $c$ with $a < c < b$ such that

$$f'(c) = 0.$$

*Proof* Let us begin by sketching a diagram (Fig. 5.3). Since $f(a) = f(b)$, the graph of $y = f(x)$ is the same height above the $x$-axis at the two ends $x = a$ and $x = b$. It seems patently obvious that there is some point $c$ between $x = a$ and $x = b$ at which the tangent is parallel to the $x$-axis, i.e. there is some point $c$ such that $f'(c) = 0$. Now let us see if we can prove this using symbols rather than geometric reasoning.

Since $f$ is continuous on the closed interval $[a, b]$, $f$ is bounded on $[a, b]$ by Theorem 4.3.1. Let

$$M = \sup_{a \leqslant x \leqslant b} f(x), \qquad m = \inf_{a \leqslant x \leqslant b} f(x).$$

and let $f(a) = f(b) = k$. Then we must have $m \leqslant k \leqslant M$.

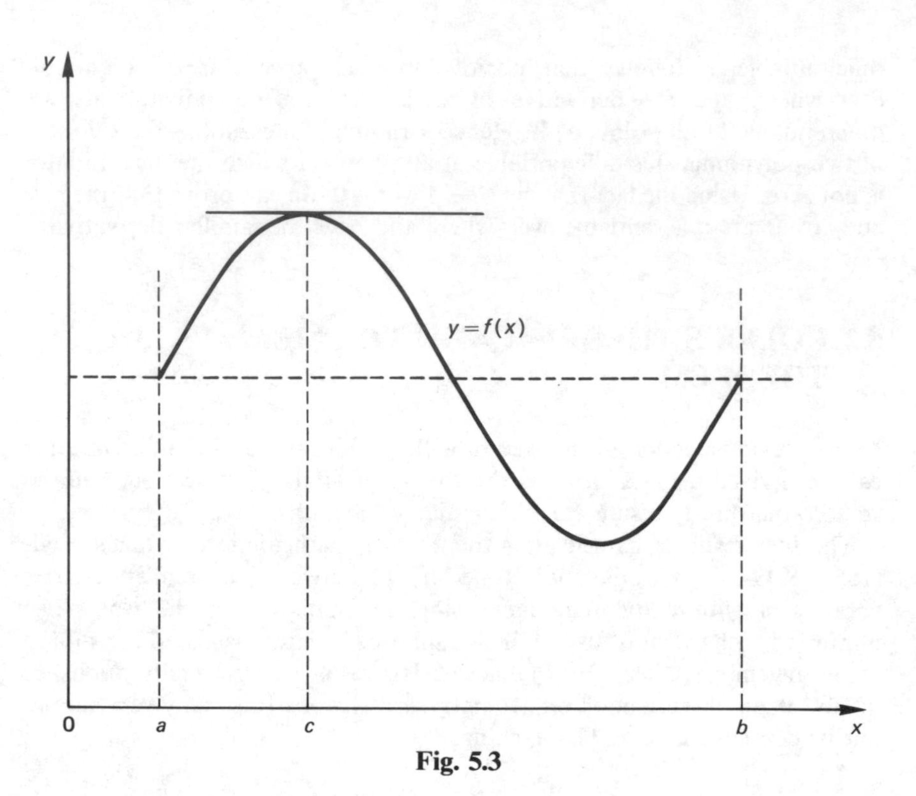


**Fig. 5.3**

*If* $m=k=M$, then $f(x)=k$ for all $x$ such that $a \leqslant x \leqslant b$. Hence $f'(c)=0$ for all $c$ such that $a<c<b$.

*If* $m \neq M$, then $m<k$ or $k<M$. Suppose $k<M$. Then there is some point $c$ with $a<c<b$ at which $f(c)=M$ by Theorem 4.3.2. Moreover, $f'(c)$ exists because $a<c<b$ by condition (ii). Now, for all $x \in [a,b]$, $f(x) \leqslant M$ and hence if $a \leqslant x < c$, then

$$\frac{f(x)-f(c)}{x-c}=\frac{f(x)-M}{x-c} \geqslant 0.$$

As $x \to c-$,
$$\frac{f(x)-f(c)}{x-c} \to f'(c)$$

and therefore
$$f'(c) \geqslant 0. \tag{1}$$

However, if $c<x \leqslant b$,

$$\frac{f(x)-f(c)}{x-c}=\frac{f(x)-M}{x-c} \leqslant 0$$

and as $x \to c+$,
$$\frac{f(x)-f(c)}{x-c} \to f'(c).$$

Hence
$$f'(c) \leqslant 0. \tag{2}$$

Since $f'(c)$ satisfies conditions (1) and (2) we must have $f'(c)=0$.

Similarly, if $m<k$, it can be proved that $f'(c_1)=0$ at the point $c_1$ at which $f(c_1)=m$. This completes the proof.

Suppose we apply this theorem in the special case in which $f(a)=f(b)=0$. Then we have a point $c$ with $a<c<b$ such that $f'(c)=0$, i.e. between any pair of zeros of $f$ there is a zero of $f'$.

*Example 5.2.1*

In the previous chapter we used the intermediate value theorem to show that the equation

$$xe^{\sin x}=\cos x$$

has a root in the interval $(0,\pi/2)$. Now we can use Rolle's theorem to prove that it does not have more than one. Write

$$f(x)=xe^{\sin x}-\cos x.$$

Then $f$ is continuous at all points of $\mathbb{R}$ and differentiable at all points of $\mathbb{R}$. If there are two distinct roots $x=a$, $x=b$ of the equation in $[0,\pi/2]$, then $0=f(a)=f(b)$ and by Rolle's theorem $f'$ is zero at some point between $a$ and $b$. However,

$$f'(x)=e^{\sin x}+x\cos xe^{\sin x}+\sin x$$

and for $0<x<\pi/2$, $f'(x)>e^0=1$. Since $f'$ does not have a zero on $(0,\pi/2)$ the equation cannot have more than one root in $(0,\pi/2)$. Thus the equation has precisely one root between 0 and $\pi/2$.

Rolle's theorem is an interesting result in its own right. It also has a part to play in proving one of the most useful results in elementary analysis—the mean value theorem.

THEOREM 5.2.2 (Mean Value Theorem) Suppose that

(i) the function $f$ is continuous on the closed interval $[a,b]$,
(ii) $f$ is differentiable on the open interval $(a,b)$.

Then there is a number $c$ with $a<c<b$ such that

$$\frac{f(b)-f(a)}{b-a}=f'(c).$$

*Proof* Again it is instructive to sketch a diagram (see Fig. 5.4). The quantity $[f(b)-f(a)]/(b-a)$ measures the gradient of the chord $AB$. The theorem

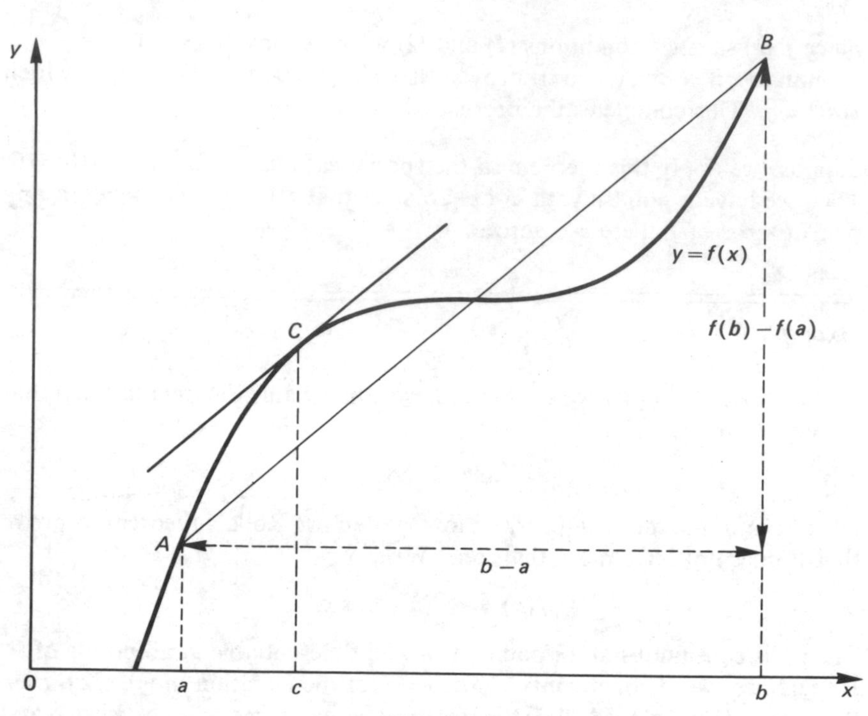


Fig. 5.4

asserts that there is a point $C$ between $A$ and $B$ with the property that the tangent at $C$ is parallel to the chord $AB$, which certainly looks a reasonable result. Indeed it can be proved rigorously.

Define a new function $g: [a,b] \to \mathbb{R}$ by the relation

$$g(x) = f(x) - kx, \tag{1}$$

where the number $k$ is chosen so that $g(a) = g(b)$

i.e. $$f(a) - ka = f(b) - kb$$

which gives $$k = \frac{f(b) - f(a)}{b - a}. \tag{2}$$

Then $g$ satisfies all the necessary conditions for the validity of Rolle's theorem. For $g$ is continuous on $[a,b]$ and $g$ is differentiable on $(a,b)$. Further, the relation $g(a) = g(b)$ is satisfied when $k$ is given by (2). Hence there is a number $c$ such that $a < c < b$ and $g'(c) = 0$ by Rolle's theorem. Now

$$g'(x) = f'(x) - k,$$

and therefore $$0 = g'(c) = f'(c) - k$$

i.e. $$f'(c) = k = \frac{f(b) - f(a)}{b - a}$$

by equation (2). This completes the proof.

Some of the applications of this result are fairly obvious. For example, whenever some estimate is required comparing the sizes of $[f(b)-f(a)]$ and $(b-a)$, the mean-value theorem is an obvious candidate. It is also used to justify the addition of an arbitrary constant when finding indefinite integrals as we will now see.

*Examples 5.2.2*

1. Suppose the function $f$ is continuous on the closed interval $[a,b]$ and differentiable on the open interval $(a,b)$. Suppose also that $f'(x)=0$ for all $x$ such that $a<x<b$. Then we can use the mean value theorem to prove that $f$ is a constant on $[a,b]$.

   Suppose $x_1$ is any number such that $a<x_1\leqslant b$. Then $f$ is continuous on $[a,x_1]$ and differentiable on $(a,x_1)$. By the mean value theorem, there is some $c$ with $a<c<x_1$ such that
$$f(x_1)-f(a)=(x_1-a)f'(c).$$
   Now $f'(x)=0$ for all $x$ such that $a<x<b$, and therefore $f'(c)=0$. Hence
$$f(x_1)-f(a)=0,$$
   i.e. $f(x_1)=f(a)$. Since this is true for all $x_1$ with $a<x_1\leqslant b$, we see that $f$ is constant on $[a,b]$.

2. Suppose the functions $f, g$ are continuous on $[a,b]$ and differentiable on $(a,b)$. Suppose also that
$$f'(x)=g'(x) \tag{1}$$
   for all $x$ such that $a<x<b$. Then the function $f-g$ is continuous on $[a,b]$ and differentiable on $(a,b)$. Moreover, $f'(x)-g'(x)=0$ for all $x$ such that $a<x<b$. Using the previous example we see that $f-g$ is constant on $[a,b]$ i.e. for all $x\in[a,b]$, $f(x)=g(x)+\text{constant}$, which explains why we add an arbitrary constant when we are finding indefinite integrals.

3. Let the function $f$ be continuous on $[a,b]$ and differentiable on $(a,b)$. Suppose that $f'(x)>0$, for all $x$ such that $a<x<b$. Then we can prove that $f$ is strictly increasing on $[a,b]$.

   Suppose $x_1, x_2$ are any two numbers such that $a\leqslant x_1<x_2\leqslant b$. Then $f$ is continuous on $[x_1,x_2]$ and differentiable on $(x_1,x_2)$. By the mean value theorem, there is a number $c$ with $x_1<c<x_2$, such that
$$f(x_2)-f(x_1)=(x_2-x_1)f'(c). \tag{1}$$
   Now $f'(c)>0$ and, therefore, equation (1) gives $f(x_2)-f(x_1)>0$, i.e. $f(x_2)>f(x_1)$. Since this is true for all $x_1, x_2$ such that $a\leqslant x_1<x_2\leqslant b$, $f$ is strictly increasing on $[a,b]$.

Finally we work an example where the solution depends on a combination of Rolle's theorem and the mean value theorem. In future we will use MVT as an abbreviation for the mean value theorem.

4. The function $f$ satisfies all the following conditions:

   (i) $f$ is continuous on the closed interval $[a,b]$;
   (ii) $f$ is twice differentiable on the open interval $(a,b)$;
   (iii) $f''(x)>0$ for all $x$ such that $a<x<b$;
   (iv) $f(a)=f(b)=0$.

   Show that $f(x)\leqslant 0$ for all $x\in[a,b]$.

*Solution* This is a fairly complicated problem and the solution is not immediately obvious. The reader may find it helpful to begin by trying to sketch a possible graph for $f$, incorporating properties (i) to (iv). Usually, very few students are lucky enough to produce a suitable picture at the first attempt. Several modifications may be needed before the sketch shows a function with all the required properties. We notice that $f''(x)>0$ for $a<x<b$ and so $f'$ is strictly increasing (see the previous example), i.e. the gradient of the tangent increases. The graph, therefore, must be drawn so that $f(a)=f(b)=0$ and the gradient of the tangent increases as $x$ increases. After a few trials the reader should come up with a picture similar to Fig. 5.5. Pictures cannot be used to prove results in analysis, but they can be used to show us the way to construct a proof. In this case the diagram seems to indicate that there is a point between $a$ and $b$ at which the tangent is parallel to the $x$-axis. Rolle's theorem shows us that this is indeed true.

Since $f''(x)$ exists for $a<x<b$, the first-order derivative $f'$ is differentiable on $(a,b)$ and so $f$ is differentiable on $(a,b)$. However, $f$ also satisfies (i) and (iv). By Rolle's theorem, therefore, there is a number $c$ with $a<c<b$ such that

$$f'(c)=0. \qquad (1)$$

Now let $a<x_1<x_2<b$. Then $f'$ is continuous on $[x_1,x_2]$ and $f'$ is differentiable on $(x_1,x_2)$. By the MVT (mean value theorem) there is some $d$, such that $x_1<d<x_2$ and

$$f'(x_2)-f'(x_1)=(x_2-x_1)f''(d)>0$$

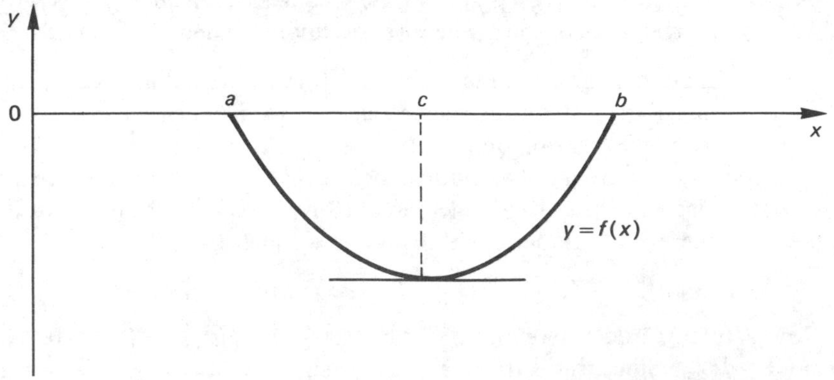

Fig. 5.5

by (iii). Thus $f'(x_2) > f'(x_1)$ and, therefore, $f'$ is strictly increasing on $(a,b)$. In particular,

$$f'(x) < 0 \qquad \text{for } a < x < c, \tag{2}$$

$$f'(x) > 0 \qquad \text{for } c < x < b, \tag{3}$$

by (1). (The geometric interpretation of relation (2) is that the tangent has a negative gradient between $a$ and $c$ and so the graph goes down, which the picture led us to expect. Relation (3) tells us that the graph goes up between $c$ and $b$, and this verifies that our sketch is not unreasonable. A further glance at the diagram also suggests that an appeal to the MVT may complete the solution.)

Again, using the MVT we can show that $f(x) \leqslant 0$ for $a \leqslant x \leqslant c$ and for $c \leqslant x \leqslant b$. First let $x$ be any number such that $a < x \leqslant c$. Then $f$ is continuous on $[a,x]$. By the MVT there is some number $c_1$ with $a < c_1 < x \leqslant c$ such that

$$f(x) - f(a) = (x-a)f'(c_1) < 0$$

by (2). Hence $f(x) < f(a) = 0$ for $a < x \leqslant c$. Similarly, if $c \leqslant x < b$, we can show that $f(x) < f(b) = 0$ by applying the MVT to $f$ on the interval $[x,b]$ and using (3). The results together give $f(x) \leqslant 0$ for $a \leqslant x \leqslant b$.

---

This long and difficult last example may have taught the reader not to despise diagrams. They certainly do not constitute a proof. They may, however, give guidance on how to set about constructing a proof and they may inspire the reader with sufficient courage to dare to tackle seemingly intractable problems.

## *EXERCISES 5.2.1*

**1** Prove that the equation $x^3 - 4x^2 + \cos x = 0$ has one and only one solution between 0 and 1.

**2** Prove that if $0 < a < b < \pi/2$, then

$$(b-a)\cos b < \sin b - \sin a < (b-a)\cos a.$$

**3** Prove that if $0 < a < b$ then

$$\left(\frac{b}{a} - 1\right) > \log\frac{b}{a} > \left(1 - \frac{a}{b}\right).$$

**4** By applying the MVT to $\sin^{-1} x$ on the interval $[0.5, 0.6]$ show that

$$\frac{\pi}{6} + \frac{\sqrt{3}}{15} < \sin^{-1}(0.6) < \frac{\pi}{6} + \frac{1}{8}$$

(Use the fact that $\sin \pi/6 = \frac{1}{2}$ i.e. $\pi/6 = \sin^{-1}\frac{1}{2}$.)

**5** The function $f$ satisfies all the following conditions:

(i) $f$ is continuous on the closed interval $[a,b]$,
(ii) $f$ is twice differentiable on the open interval $(a,b)$;

(iii) $f''(x)<0$ for all $x$ such that $a<x<b$;
(iv) $f(a)=f(b)=0$.

Show that $f(x)\geqslant 0$ for all $x\in[a,b]$.

6 Show that if $0<a<b<1$ then

$$\frac{b-a}{\sqrt{1-a^2}}<\sin^{-1}b-\sin^{-1}a<\frac{b-a}{\sqrt{1-b^2}}.$$

7 The function $f:\mathbb{R}\to\mathbb{R}$ is differentiable at all points of $\mathbb{R}$, and $f(0)>0$ and $f'(x)<\frac{1}{2}$ for all $x$ such that $x\geqslant 0$. By considering the function given by $g(x)=f(x)-x$, show there is one and only one positive real number $c$ such that $f(c)=c$.

## 5.3 TAYLOR'S THEOREM

The conclusions of the MVT can be presented as

$$f(b)=f(a)+(b-a)f'(c). \tag{1}$$

This relation suggests that it might be possible to extend the result to include higher-order derivatives. Under suitable conditions, the corresponding formula would appear to be

$$f(b)=f(a)+f'(a)(b-a)+\frac{f''(a)}{2!}(b-a)^2+\ldots$$
$$+\frac{f^{(n-1)}}{(n-1)!}(b-a)^{n-1}+\frac{f^{(n)}(c)}{n!}(b-a)^n,$$

where $a<c<b$. This result is sometimes known as the $n$th mean value theorem or it is called Taylor's theorem. Many proofs exist in the literature and some of them are amazingly slick and somewhat bewildering at first sight. We will give one of the longer proofs which has the merit of being relatively easy to understand.

THEOREM 5.3.1 (Taylor's Theorem; $n$th MVT)

Suppose that $f$ possesses derivatives of all orders up to and including the $(n-1)$th-order derivative on the closed interval $[a,b]$. If

(i) $f$ and its first $(n-1)$ derivatives are continuous on the closed interval $[a,b]$;
(ii) $f$ has an $n$th-order derivative on the open interval $(a,b)$;

then there is some number $c$ such that $a<c<b$ and

$$f(b)=f(a)+f'(a)(b-a)+\frac{f''(a)}{2!}(b-a)^2+\dots$$
$$+\frac{f^{(n-1)}(a)}{(n-1)!}(b-a)^{n-1}+\frac{f^{(n)}(c)}{n!}(b-a)^n.$$

*Proof* We proceed in a similar fashion to the proof of the MVT. First construct a suitable function $g$ and then apply Rolle's theorem. In this case Rolle's theorem has to be used repeatedly.

Define the function $g$ by

$$g(x)=f(x)-f(a)-f'(a)(x-a)-\frac{f''(a)}{2!}(x-a)^2-\dots$$
$$-\frac{f^{(n-1)}(a)}{(n-1)!}(x-a)^{n-1}-k(x-a)^n,$$
$$=f(x)-f(a)-\sum_{r=1}^{n-1}\frac{f^{(r)}(a)}{r!}(x-a)^r-k(x-a)^n, \tag{1}$$

where the number $k$ is chosen so that $g(b)=g(a)$. The value of $k$ is therefore

$$k=\frac{f(b)-f(a)-\sum_{r=1}^{n-1}\frac{f^{(r)}(a)}{r!}(b-a)^r}{(b-a)^n}. \tag{2}$$

Now the function $g$ is continuous on $[a,b]$ and differentiable on $(a,b)$ and $g(a)=g(b)$. By Rolle's theorem there is some number $c_1$ with $a<c_1<b$ such that

$$g'(c_1)=0. \tag{3}$$

Now differentiating (1) we have

$$g'(x)=f'(x)-f'(a)-f''(a)(x-a)-\dots-\frac{f^{(n-1)}(a)}{(n-2)!}(x-a)^{n-2}-kn(x-a)^{n-1},$$

and therefore $g'(a)=0$. Assuming that $n\geqslant 2$, we see that the function $g'$ is continuous on $[a,c_1]$ and it is differentiable on $(a,c_1)$, and $g'(a)=g'(c_1)=0$. By Rolle's theorem, there is a number $c_2$ such that $a<c_2<c_1<b$ and

$$g''(c_2)=0. \tag{4}$$

We can check, by differentiation, that

$$g^{(k)}(a)=0 \qquad (1\leqslant k\leqslant n-1).$$

In particular, if $n\geqslant 3$, then $g''(a)=0=g''(c_2)$ and applying Rolle's theorem to $g''$ on $[a,c_2]$ we obtain a number $c_3$ such that $a<c_3<c_2<c_1<b$ and $g'''(c_3)=0$. Continue in this way, and repeat the process a further $n-3$ times, to obtain a number $c_n$ such that

$$a<c_n<c_{n-1}<\dots<c_3<c_2<c_1<b$$

and $$g^{(n)}(c_n) = 0. \tag{5}$$

From (1), $$g^{(n)}(x) = f^{(n)}(x) - kn! \qquad (a < x < b),$$

and therefore equation (5) gives (writing $c$ in place of $c_n$)

$$f^{(n)}(c) = n!k.$$

When this is combined with equation (2), we obtain the required result, and the proof is complete.

There are dozens of practical applications of Taylor's theorem as we will see a little later. Some of them will use Taylor's theorem in precisely the form in which it is stated in Theorem 5.3.1. Others use very slight variants of it. For example, Theorem 5.3.1 is stated for the case $a < b$. With a very slight modification of the proof and an appropriate change to the intervals, it remains true when $b < a$. In this case the result is stated in the following way.

THEOREM 5.3.1(a) Let $a$ be two real numbers with $b < a$. Suppose $f$ possesses derivatives of all orders up to and including the $(n-1)$th order derivative on the closed interval $[b, a]$. If

(a) $f$ and its first $(n-1)$ derivatives are continuous on the closed interval $[b, a]$, and
(b) $f$ has an $n$th order derivative on the open interval $(b, a)$,

then there is some number $c$ such that $b < c < a$ and

$$f(b) = f(a) + f'(a)(b-a) + \frac{f''(a)}{2!}(b-a)^2 + \dots$$
$$+ \frac{f^{(n-1)}(a)}{(n-1)!}(b-a)^{n-1} + \frac{f^{(n)}(c)}{n!}(b-a)^n.$$

The proof uses exactly the same function $g$ as before. In this case the only slight modification is that we apply Rolle's theorem to intervals $[b, a]$, $[c_1, a]$, $[c_2, a]$, etc. instead of the intervals $[a, b]$, $[a, c_1]$, $[a, c_2]$.

Our first application, commonly known as l'Hôpital's rule (or l'Hospital's rule) is named after a French mathematician, the Marquis G. F. A. de l'Hôpital (1661–1704). The spelling is somewhat doubtful—sometimes the letter 's' is included, at other times the 's' is omitted and a circumflex placed over the letter 'o'. There is, however, little doubt as to the identity of the discoverer of the rule, and it wasn't the Marquis! For many years it has been known that the result was originally discovered by the Swiss mathematician Jean Bernouilli (1667–1748), a member of the famous Bernouilli family which produced a remarkable number of celebrated mathematians over the years. The story begins in 1692, when the two men met in Paris. Bernouilli had already spent some time studying Leibniz's work on calculus and he taught it to l'Hôpital. But what was more significant is that they signed an agreement. According to this contract, the Marquis guaranteed to pay Bernouilli a regular salary and Bernouilli undertook to send all his mathematical discoveries to l'Hôpital. The Marquis was then free to use them in any way he pleased.

In 1696, the Marquis de l'Hôpital published a text book on calculus. It included material which had been communicated by Bernouilli and among this material was the law which we now know as l'Hôpital's rule. The book was well written and its author proved to be a good teacher. The preface includes an acknowledgement of the importance of the work of Leibniz and Bernouilli. To this day, however, the Marquis still gets the credit, in name at least, for the following law.

## 5.4 L'HÔPITAL'S RULE

In Chapter 4 we saw that if $f(x) \to l$ and $g(x) \to m$ as $x \to a$, then $f(x)/g(x) \to l/m$ as $x \to a$ provided $m \neq 0$ and $g(x) \neq 0$ sufficiently close to $a$. If $l = 0$ and $m = 0$, then the algebra of limits is not applicable. This is just the case in which l'Hôpital's rule may be applicable.

THEOREM 5.4.1 (L'Hôpital's Rule) Suppose there is some positive real number $R$ such that the functions $f$ and $g$ and their first $n$ derivatives are continuous on the open interval $(a - R, a + R)$, where $n \geqslant 1$. Suppose also that

(i) $f(a) = 0$, $f^{(k)}(a) = 0$ for $k \leqslant n - 1$,
(ii) $g(a) = 0$, $g^{(k)}(a) = 0$ for $k \leqslant n - 1$,
(iii) $g^{(n)}(a) \neq 0$.

Then
$$\frac{f(x)}{g(x)} \to \frac{f^{(n)}(a)}{g^{(n)}(a)}$$
as $x \to a$.

*Proof* Since $g^{(n)}$ is continuous on $(a - R,\ a + R)$ and $g^{(n)}(a) \neq 0$, it follows from Theorem 4.2.1 that there is some $r > 0$ (with $r \leqslant R$) such that $g^{(n)}(t) \neq 0$ for all $t$ such that $a - r < t < a + r$.

Now let $x$ be any real number such that $0 < |x - a| < r$, then $f, g$ and their first $n$ derivatives are continuous on the closed interval with end-points $a, x$. By Taylor's theorem,

$$f(x) = f(a) + f'(a)(x - a) + \ldots + \frac{f^{n-1}(a)}{(n-1)!}(x - a)^{n-1} + \frac{f^{(n)}(c)}{n!}(x - a)^n$$
$$= \frac{f^{(n)}(c)}{n!}(x - a)^n \tag{1}$$

by condition (i), where $c$ lies between $a$ and $x$. Similarly,

$$g(x)=g(a)+g'(a)(x-a)+\ldots+\frac{g^{(n-1)}(a)}{(n-1)!}(x-a)^{n-1}+\frac{g^{(n)}(d)}{n!}(x-a)^n$$
$$=\frac{g^{(n)}(d)}{n!}(x-a)^n \qquad (2)$$

by condition (ii), where $d$ lies between $a$ and $x$. Hence $0<|a-d|<r$, because $0<|a-x|<r$. Since $g^{(n)}(t)\neq 0$ for all $t$ such that $a-r<t<a+r$, we see $g^{(n)}(d)\neq 0$ and so $g(x)\neq 0$ provided $0<|x-a|<r$. Thus, if $0<|x-a|<r$, then

$$\frac{f(x)}{g(x)}=\frac{f^{(n)}(c)}{g^{(n)}(d)} \qquad (3)$$

from (1) and (2). As $x\to a$, we must have $c\to a$ and $d\to a$, because $c, d$ are between $a$ and $x$. Moreover, $f^{(n)}(c)\to f^{(n)}(a)$ as $c\to a$ and $g^{(n)}(d)\to g^{(n)}(a)$ as $d\to a$, because $f^{(n)}, g^{(n)}$ are continuous. Hence

$$\frac{f^{(n)}(c)}{g^{(n)}(d)}\to\frac{f^{(n)}(a)}{g^{(n)}(a)}$$

as $x\to a$, and therefore, from (3),

$$\frac{f(x)}{g(x)}\to\frac{f^{(n)}(a)}{g^{(n)}(a)}$$

as $x\to a$. This completes the proof.

Taylor's theorem states that a number $c$ exists with given properties when $f$ satisfies certain conditions. The number $c$ is determined by the function $f$ and the interval $[a,b]$; it is not just an arbitrary number between $a$ and $b$. It was therefore necessary in the above proof to use two different symbols $c, d$ for the intermediate points—one for the function $f$ and the other for $g$.

Now let us test how this rule works in practice by trying an example.

### *Examples 5.4.1*

Evaluate the following limits:

1. $\lim_{x\to 0}\frac{e^{x^2}-1}{\sin x^2}$. 2. $\lim_{x\to 0}\frac{\cos x-1}{x^2}$. 3. $\lim_{x\to 1}\frac{(x-1)^3}{\log x}$. 4. $\lim_{x\to 0}\frac{\sin x}{\cosh x}$.

*Solutions*

1. We notice that if

$$f(x)=e^{x^2}-1,\ g(x)=\sin x^2$$

for all real $x$, then $f(0)=0$, $g(0)=0$. So we have an obvious candidate for a trial of l'Hôpital's rule. Obviously, we must check that all the conditions are satisfied before applying it. We must, therefore, decide what value to use for $n$ and we must check that there is some suitable $R$. How do we find $n$? Theorem 5.4.1 says that $n$ is the smallest positive integer for which the $n$th-order derivative of $g$ is non-zero. We must, therefore, find some derivatives of $g$. Now $g'(x)=2x\cos x^2$ and $g'(0)=0$,

$$g''(x)=-4x^2\sin x^2+2\cos x^2 \qquad \text{and} \qquad g''(0)=2\neq 0.$$

The first non-zero derivative at the origin is the second and we therefore use $n=2$. Naturally, we still have to check the conditions on $f$. We notice that

$$f'(x)=2xe^{x^2} \qquad \text{and} \qquad f'(0)=0,$$

$$f''(x)=4x^2e^{x^2}+2e^{x^2}.$$

Moreover, $f, g$ and all their derivatives are continuous at all points of $\mathbb{R}$, and we may therefore choose any positive value for $R$. Suppose we choose $R=1$. Clearly $f, g, f', g', f''$ and $g''$ are continuous on $(-1,+1)$. Further, $f(0)=f'(0)=0$, and $g(0)=g'(0)=0$, $g''(0)\neq 0$. By l'Hôpital's rule

$$\lim_{x\to 0}\frac{f(x)}{g(x)}=\frac{f''(0)}{g''(0)}=\frac{2}{2}=1,$$

i.e.

$$\lim_{x\to 0}\frac{e^{x^2}-1}{\sin x^2}=1.$$

For most undergraduates, one of the main problems is what value to use for $n$. Students expect somehow to look at the question and immediately know which value to choose, but in practice this is impossible. The functions $f$ and $g$ are usually fairly obvious. After that it is a question of finding the derivatives and evaluating them at the appropriate point. The integer $n$ is the order of the lowest-order non-zero derivative of $g$. The reader should experience no difficulty whatsoever in finding $n$, when this method is followed. Finally, a word of warning. Don't use l'Hôpital's rule if it is not appropriate. If $f(a)\neq 0$, $g(a)\neq 0$ then l'Hôpital's rule is not appropriate and should not be used.

2. We follow the same method as before.

Let

$$f(x)=\cos x-1,\ g(x)=x^2$$

for all $x$. Then

$$g'(x)=2x \qquad \text{and} \qquad g'(0)=0,$$

$$g''(x)=2 \qquad \text{and} \qquad g''(0)=2\neq 0$$

and we again have a problem for which $n=2$. Moreover,

$$f'(x)=-\sin x \qquad \text{and} \qquad f'(0)=0,$$

$$f''(x)=-\cos x.$$

In this case $f, g$ and all their derivatives are continuous at every point of $\mathbb{R}$ and we can choose any positive value for $R$. For example, we can use $R=1$, since $f, g, f', g', f''$ and $g''$ are all continuous on $(-1,1)$. Hence, using l'Hôpital's rule,

$$\lim_{x-0}\frac{\cos x-1}{x^2}=\frac{f''(0)}{g''(0)}=\frac{-1}{2}=-\frac{1}{2}.$$

3. By now, the reader is sufficiently familiar with the method to be able to write down the solution quite quickly. There are, however, one or two details which

require just a little care. The choice of the functions $f, g$ is fairly obvious. Let

$$f(x) = (x-1)^3 \qquad (\text{all } x \in \mathbb{R})$$
$$g(x) = \log x \qquad (x > 0)$$

Then
$$f(1) = 0, \qquad g(1) = 0$$
$$g'(x) = 1/x \qquad (x > 0),$$

and
$$g'(1) = 1.$$

This problem seems an obvious candidate for the use of l'Hôpital's rule with $n = 1$ for this particular example. Since we require a limit as $x$ tends to 1, the continuity conditions must be checked on an interval about 1. In this case the interval $(0, 2)$ will do. Clearly, $f, g, f', g'$ are continuous on $(0, 2)$. By l'Hôpital's rule,

$$\lim_{x \to 1} \frac{(x-1)^3}{\log x} = \frac{f'(1)}{g'(1)} = \frac{0}{1} = 0.$$

In this example $n = 1$ and $f'(1) = 0$. This causes no problem, as the conditions for l'Hôpital's rule require $g^{(n)}(a) \neq 0$; they do **not** stipulate that $f^{(n)}(a)$ is non-zero.

One final word of warning about this example—the question asks for the limit as $x$ tends to 1 and so $f, g$ and their derivatives are evaluated at 1. When the question asks for a limit as $x$ tends to a non-zero value, then answer the question you were asked—don't evaluate the functions at zero and then wonder why your answer is wrong.

4. Use
$$f(x) = \sin x,$$
$$g(x) = \cosh x$$
for all real $x$. Then $f(0) = 0$, $g(0) = 1 \neq 0$, and this is a case in which l'Hôpital's rule is **not** appropriate. In fact, we can see immediately that $f(x) \to 0$ as $x \to 0$, $g(x) \to 1$ as $x \to 0$ and so
$$\frac{f(x)}{g(x)} \to \frac{0}{1}$$
as $x \to 0$ by the algebra of limits.

## *EXERCISES 5.4.1*

Decide which of the following limits exist. When the limit exists find its value.

**1** $\displaystyle\lim_{x \to 1} \frac{\log x^2}{1-x}$, **2** $\displaystyle\lim_{x \to 1} \frac{(x^2-1)^2}{1-\sin(\pi x/2)}$,

**3** $\displaystyle\lim_{x \to 1} \frac{\cos x}{\sin x}$, **4** $\displaystyle\lim_{x \to 1} \frac{\cos x}{1-x}$,

**5** $\displaystyle\lim_{x \to 1} \frac{(1-x)^2}{\sin^2(\pi x)}$, **6** $\displaystyle\lim_{x \to 1} \frac{(x-1)^3}{\log x}$,

**7** $\displaystyle\lim_{x \to 0} \frac{1-\log(e+x^2)}{(e^x-1)\sin x}$, **8** $\displaystyle\lim_{x \to 1} \frac{\log x}{x^2-1}$,

**9** $\lim_{x\to 2}\dfrac{x-2}{x^2-4}$, **10** $\lim_{x\to 0}\dfrac{1-\cosh x}{x^2}$,

**11** $\lim_{x\to 0}\dfrac{\sinh x^2}{\sin x^2}$, **12** $\lim_{x\to 1}\dfrac{\sin \pi x}{\log x}$,

**13** $\lim_{x\to 0}\dfrac{\log(\cos x)}{x^2}$, **14** $\lim_{x\to 2}\dfrac{x^3-x^2-x-2}{x^3-3x^2+3x-2}$,

**15** $\lim_{x\to 0}\dfrac{e^{\pi x}-e^{-\pi x}}{\log(1+\pi x)}$, **16** $\lim_{x\to \pi/2}\dfrac{\cosh x}{\sin x}$.

(Don't use l'Hôpital's rule when it is not appropriate. You have been warned!)

## 5.5 MAXIMA AND MINIMA

Most sixth-formers who study mathematics spend at least some time working on problems on maxima and minima. Given a reasonably well-behaved function $f$ such that $f'(a)=0$, $f''(a)>0$, the conclusion is that $f$ has a minimum at $a$. If $f'(a)=0$ and $f''(a)<0$, then there is a maximum at $a$. But what happens if $f'(a)=0$ and $f''(a)=0$? Among sixth-formers brave enough to hazard any guess at all many jump to the conclusion that $f$ has a point of inflection at $a$. Is this really true? Perhaps we were a little hasty, for if $f(x)=(x-a)^4$, then $f'(a)=0, f''(a)=0$. However, $f(x)=(x-a)^4>0=f(a)$ for all $x$ such that $x\neq a$ and so it would appear that $f$ has a minimum at $a$ despite the fact that $f''(a)=0$. This suggests that it might be profitable to look into the subject in a little more detail. Perhaps we should begin by spelling out carefully the meaning of the terms we intend to use. First of all we introduce the terms local maximum and local minimum.

DEFINITION 5.5.1 Suppose there is some $R>0$ such that $f(x)$ is defined and satisfies the inequality $f(x)\leqslant f(a)$ for all $x$ such that $a-R<x<a+R$. Then $f$ is said to have a **local maximum** at $a$.

DEFINITION 5.5.2 Suppose there is some $R>0$ such that $f(x)$ is defined and satisfies the inequality $f(x)\geqslant f(a)$ for all $x$ such that $a-R<x<a+R$. Then $f$ is said to have a **local minimum** at $a$.

If $f$ has a local maximum at $a$, then $f(a)$ is the greatest value assumed by $f(x)$ for $a-R<x<a+R$ for some $R>0$, i.e. $f(a)$ is the greatest value assumed by $f(x)$ for $x$ close to $a$. For this reason we call it a **local** maximum. There may, of course, be points outside the interval $(a-R,\ a+R)$ at which $f(x)$

assumes values greater than $f(a)$. In contrast, we have the notion of a global maximum. Let $f: [a,b] \to \mathbb{R}$ have the property that there is some $c \in [a,b]$ such that $f(x) \leqslant f(c)$ for all $x \in [a,b]$. Then $f(c)$ is the largest value assumed by $f(x)$ on $[a,b]$. Thus $f(c)$ is a global maximum rather than a mere local phenomenon. Similarly, if $d \in [a,b]$ is such that $f(x) \geqslant f(d)$ for all $x \in [a,b]$, then $f(d)$ is the smallest value assumed by $f(x)$ on $[a,b]$ and it is, therefore, a global minimum. Now let us begin the search for the conditions needed for a maximum or minimum.

THEOREM 5.5.1 Suppose $f(x)$ is defined for all $x$ such that $a-R<x<a+R$, where $R$ is a given positive real number. If $f$ has a local maximum (or minimum) at $a$ and if $f'(a)$ exists, then

$$f'(a)=0.$$

*Proof* Since $f$ has a local maximum at $a$, there is some $r>0$ (with $r \leqslant R$) such that $f(x) \leqslant f(a)$ for all $x$ such that $a-r<x<a+r$.
Thus if $a-r<x<a$ then

$$\frac{f(x)-f(a)}{x-a} \geqslant 0. \tag{1}$$

Now, $f'(a)$ exists and so

$$f'(a) = \lim_{x \to a-} \frac{f(x)-f(a)}{x-a} \geqslant 0 \tag{2}$$

by relation (1). Similarly, if $a<x<a+r$, then

$$\frac{f(x)-f(a)}{x-a} \leqslant 0$$

and

$$f'(a) = \lim_{x \to a+} \frac{f(x)-f(a)}{x-a} \leqslant 0. \tag{3}$$

From (2) and (3) we see that $f'(a)=0$.

A similar proof shows that if $f$ has a local minimum at $a$, then $f'(a)=0$. This completes the proof.

DEFINITION 5.5.3 Suppose that $f(x)$ is defined for all $x$ such that $a-R<x<a+R$, for some positive real number $R$, and suppose also that $f'(a)$ exists. If $f'(a)=0$ then $a$ is called a **stationary point** or **critical point**.

Now, how do we set about distinguishing the characteristics of our stationary points? Probably our wisest course of action is to use Taylor's theorem.

Suppose $f$ and all its first $n$ derivatives ($n \geqslant 2$) are continuous on some interval $(a-R, a+R)$, where $R$ is some given positive real number. Suppose also that $f'(a)=0$ so that $a$ is a stationary point of $f$. If $f^{(k)}(a)=0$ for $1 \leqslant k \leqslant n-1$ and $f^{(n)}(a) \neq 0$, then Taylor's theorem gives us

$$f(x)=f(a)+f'(a)(x-a)+\ldots+\frac{f^{(n-1)}(a)}{(n-1)!}(x-a)^{n-1}+\frac{f^{(n)}(c)}{n!}(x-a)^n$$
$$=f(a)+\frac{f^{(n)}(c)}{n!}(x-a)^n$$

for $0<|x-a|<R$, where $c$ lies between $a$ and $x$. Now let us look at some separate cases.

(i) *Suppose $n$ is even ($n \geqslant 2$) and $f^{(n)}(a)>0$.* Since $f^{(n)}$ is continuous on $(a-R, a+R)$ there is some $r>0$ (with $r \leqslant R$) such that $f^{(n)}(t)>0$ for all $t$ such that $a-r<t<a+r$. Thus for all $x$ with $0<|x-a|<r$, we have

$$f(x)-f(a)=\frac{f^{(n)}(c)}{n!}(x-a)^n>0.$$

Since $f^{(n)}(c)>0$ because $c$ lies between $a$ and $x$. Hence

$$f(x)>f(a)$$

for all $x$ such that $0<|x-a|<r$ and $f$ has a local minimum at $a$.

(ii) *Suppose $n$ is even ($n \geqslant 2$) and $f^{(n)}(a)<0$.* As before, the continuity of $f^{(n)}$ means that there is some $r>0$ (with $r \leqslant R$) such that $f^{(n)}(t)<0$ for all $t$ such that $a-r<t<a+r$. Thus, for all $x$ with $0<|x-a|<r$, we have

$$f(x)-f(a)=\frac{f^{(n)}(c)}{n!}(x-a)^n<0,$$

i.e.

$$f(x)<f(a)$$

and $f$ has a local maximum at $a$.

(iii) *Suppose $n$ is odd ($n \geqslant 2$) and $f^{(n)}(a)>0$.* We use the same $r$ as in (i). Then for $a<x<a+r$,

$$f(x)-f(a)=\frac{f^{(n)}(c)}{n!}(x-a)^n>0,$$

and for $a-r<x<a$,

$$f(x)-f(a)=\frac{f^{(n)}(c)}{n!}(x-a)^n<0,$$

since $f^{(n)}(c)>0$ because $c$ lies between $a$ and $x$. Thus

$$f(x)>f(a)\ (a<x<a+r)$$
$$f(x)<f(a)\ (a-r<x<a)$$

and $f$ has neither a local maximum nor a local minimum at $a$. Such a point is called a point of inflection.

(iv) *Suppose $n$ is odd ($n \geq 2$) and $f^{(n)}(a) < 0$.* We use the same $r > 0$ as in (ii). If we follow the method used in (iii) then we see that

$$f(x) < f(a) \qquad (a < x < a + r),$$

$$f(x) > f(a) \qquad (a - r < x < a).$$

Again there is neither a local maximum nor a local minimum at $a$, and $f$ has a point of inflection at $a$.

Let us gather together these results and incorporate them in a theorem.

THEOREM 5.5.2 Suppose $f$ and its first $n$ derivatives ($n \geq 2$) are continuous on the interval $(a - R, a + R)$, where $R$ is some given positive real number. Suppose also that $f^{(k)}(a) = 0$ ($1 \leq k \leq n - 1$) and $f^{(n)}(a) \neq 0$. If

(i) $f^{(n)}(a) > 0$ and $n$ is an even integer, then $f$ has a local minimum at $a$;
(ii) $f^{(n)}(a) < 0$ and $n$ is an even integer, then $f$ has a local maximum at $a$;
(iii) $n$ is an odd integer, then $f$ has a point of inflection at $a$.

The special case $n = 2$ is the one normally encountered in the sixth form. From our theorem we see that if $f'(a) = 0$, $f''(a) > 0$, then we have a local minimum. However, if $f'(a) = 0$ and $f''(a) < 0$ then we have a local maximum. These results are, of course, very familiar already. However, with a well-behaved function the case $f'(a) = 0$, $f''(a) = 0$ causes no problem, now that our general theorem is available. All we have to do in such cases is to keep on differentiating until we reach a derivative which is non-zero at $a$. Let us look at some particular examples.

---

*Examples 5.5.1*

1. Let $f(x) = (x - a)^4$. Then $f$ and all its derivatives are continuous at every point of $\mathbb{R}$ and so any positive number can be used for $R$. Moreover, $f'(a) = 0$, $f''(a) = 0$, $f'''(a) = 0$, $f^{(4)}(a) = 4!$ This function satisfies all the conditions of the theorem with $n = 4$ and $f^{(n)}(a) = 4! = 24 > 0$. Thus $f$ has a local minimum at $a$ (see Theorem 5.5.2(i)), which is precisely what we conjectured earlier.

2. Let $f(x) = \sin x - x$. Then $f$ and all its derivatives are continuous at every point of $\mathbb{R}$. Now,

$$f'(x) = \cos x - 1,$$

$$f''(x) = -\sin x,$$

$$f'''(x) = -\cos x,$$

and, for every integer $k$,

$$f'(2k\pi)=0, \qquad f''(2k\pi)=0, \qquad f'''(2k\pi)=-1.$$

We therefore need to use Theorem 5.5.2(iii). Since the appropriate value for $n$ is $n=3$, this shows us that $f$ has a point of inflection at each of the points $2k\pi$ ($k=0, \pm 1, \pm 2, \pm 3, \ldots$). In this example, all the stationary points are points of inflection.

Suppose now that we have a function $f$ which is continuous on a closed interval $[a,b]$. Then Theorem 4.3.1 and 4.3.2 tell us that $f(x)$ has a largest value $M$ on $[a,b]$. How do we set about finding this largest value? There are two possibilities: either $f(a)=M$ or $f(b)=M$ or there is a point $c$ with $a<c<b$ such that $f(c)=M$. If $f(c)=M$, where $a<c<b$, then $f(x) \leqslant f(c)=M$ for all $x$ such that $c-R<x<c+R$ where $R$ is the smaller of the two numbers $c-a$, $b-c$. In this case, therefore, $f$ has a local maximum at $c$. We already know that, if $f'(c)$ exists and $f$ has a local maximum at $c$, then $f'(c)=0$. A similar result holds for the minimum value of $f(x)$ on $[a,b]$.

3. The function $f\colon[0,5] \to \mathbb{R}$ is given by

$$f(x)=x^3-5x^2+3x+23.$$

Find the largest and smallest values assumed by $f(x)$ on $[0,5]$.

*Solution* This function $f$ is differentiable on $(0,5)$. Hence the largest value of $f(x)$ will occur either at one of the end-points or at a point in between at which there is a local maximum, i.e. it will either occur at an end-point or at an intermediate point at which the derivative vanishes. A similar result holds for the smallest value. We therefore need only find the points $c$ with $0<c<5$ at which $f'(c)=0$ and evaluate $f(x)$ at such points $c$ and at the end-points 0 and 5.

Now, $$f'(x)=3x^2-10x+3=(3x-1)(x-3)=0,$$

if $x=\frac{1}{3}$, 3. It is easy to check that $f(0)=23$, $f(\frac{1}{3})=23\frac{13}{27}$, $f(3)=14$, $f(5)=48$. We can now simply read off the result. The largest value of $f(x)$ on the interval is 48, the smallest value is 14.

---

Finally we use Taylor's theorem to develop power series expansions.

### 5.5.1 Taylor series

Let $a, x$ be two distinct real numbers and let $J$ be the closed interval with end-points $a, x$. Suppose that the function $f$ possesses continuous derivatives of all orders on $J$. Then by Taylor's theorem we have, for all $n \geqslant 1$,

$$f(x)=f(a)+f'(a)(x-a)+\ldots+\frac{f^{(n-1)}(a)}{(n-1)!}(x-a)^{n-1}+\frac{f^{(n)}(c)}{n!}(x-a)^n$$

$$=\sum_{k=0}^{n-1} a_k(x-a)^k+\frac{f^{(n)}(c)}{n!}(x-a)^n,$$

where $a_0=f(a)$, $a_k=f^{(k)}(a)/k!$ ($k=1,\ldots,n-1$) and $c$ lies between $a$ and $x$.

Thus $$f(x)-\sum_{k=0}^{n-1} a_k(x-a)^k=\frac{f^{(n)}(c)}{n!}(x-a)^n;$$

and, if we can show (for a particular value of $x$) that

$$\frac{f^{(n)}(c)}{n!}(x-a)^n$$

tends to zero as $n\to\infty$, then (for this value of $x$) the infinite series $\sum_{k=0}^{\infty} a_k(x-a)^k$ converges and its sum is $f(x)$. We can then write

$$f(x)=\sum_{k=0}^{\infty} a_k(x-a)^k$$

where $a_0=f(a)$, $a_k=f^{(k)}(a)/k!$ $(k\geqslant 1)$. This series is called the Taylor series of $f(x)$ about $x=a$. Let us try a few examples.

*Examples 5.5.2*

1. Let $f(x)=\mathrm{e}^x$ for all $x$. Then $f$ and all its derivatives are continuous at all points of $\mathbb{R}$. Let $x$ be any non-zero real number and apply Taylor's theorem on the closed interval $J$ with end-points 0 and $x$ to obtain

$$f(x)=f(0)+f'(0)x+\frac{f''(0)}{2!}x^2+\ldots+\frac{f^{(n-1)}(0)}{(n-1)!}x^{n-1}+\frac{f^{(n)}(c)}{n!}x^n, \quad (1)$$

where $c$ lies between 0 and $x$. Now we can easily check that

$$f(0)=1, \quad f^{(k)}(0)=1$$

for all positive integers $k$. Hence (1) reduces to

$$f(x)=1+x+\frac{x^2}{2!}+\frac{x^3}{3!}+\ldots+\frac{x^{n-1}}{(n-1)!}+\frac{\mathrm{e}^c}{n!}x^n. \quad (2)$$

Now $c$ lies between 0 and $x$ and therefore

$$\left|\frac{\mathrm{e}^c x^n}{n!}\right|\leqslant\frac{\mathrm{e}^{|x|}|x|^n}{n!}.$$

For any given fixed real number $x$,

$$\frac{\mathrm{e}^{|x|}|x|^n}{n!}\to 0$$

as $n\to\infty$. Thus for all real numbers $x$

$$f(x)=\mathrm{e}^x=1+x+\frac{x^2}{2!}+\frac{x^3}{3!}+\ldots=\sum_{n=0}^{\infty}\frac{x^n}{n!}. \quad (3)$$

This expansion can be used to obtain a rather neat limit.

Let $\alpha$ be any real number and choose a positive integer $q$ such that $q > \alpha$. Then, for $x > 1$, we have

$$0 < x^{\alpha} < x^{q}. \tag{4}$$

Now, as $x \to \infty$, $x^{q} \to \infty$ and $e^{x} \to \infty$. What happens to the quotient $x^{q}/e^{x}$? Relation (3) can be used to give an answer. For, if $x > 1$, then

$$e^{x} > x^{q+1}/(q+1)!$$

from (3) and combining this with (4) yields the inequality

$$0 < \frac{x^{\alpha}}{e^{x}} < \frac{x^{q}}{e^{x}} < \frac{x^{q}(q+1)!}{x^{q+1}} = \frac{(q+1)!}{x}$$

i.e.
$$0 < x^{\alpha}c^{-x} < (q+1)!/x \qquad (x > 1).$$

As $x \to \infty$, $(q+1)!/x \to 0$, and therefore, using the sandwich rule,

$$\boxed{x^{\alpha}e^{-x} \to 0 \text{ as } x \to \infty.} \tag{5}$$

It is therefore clear that as far as this limit is concerned, the exponential dominates the power. The reader will do well to make an effort to memorise the result. The corresponding rule for logarithms is that for all $\beta > 0$

$$\boxed{\frac{\log y}{y^{\beta}} \to 0 \text{ as } y \to \infty.}$$

This can be quite easily deduced from (5). For using $\alpha = 1$ we see that $xe^{-x} \to 0$ as $x \to \infty$. Now write $x = \beta t$ ($\beta > 0$). As $x \to \infty$, $t \to \infty$ and

$$\beta t e^{-\beta t} \to 0$$

and therefore
$$t e^{-\beta t} \to 0$$

as $t \to \infty$. Now write $t = \log y$, so that $e^{t} = y$ and the relation becomes

$$\frac{\log y}{y^{\beta}} \to 0$$

as $y \to \infty$, for $\beta > 0$.

2. Let $f(x) = \log(1+x)$ $(x > -1)$, and use $a = 0$ in Taylor's theorem. First let $x$ be any real number such that $0 < |x| < 1$. Then $f$ and its derivatives of all orders are continuous on the closed interval with end-points 0 and $x$. Using Taylor's theorem,

$$f(x) = f(0) + f'(0)x + \frac{f''(0)}{2!}x^{2} + \ldots + \frac{f^{(n-1)}(0)}{(n-1)!}x^{n-1} + \frac{f^{(n)}(c)}{n!}x^{n},$$

where $c$ lies between 0 and $x$. By differentiating, we find that

$$f^{(k)}(t)=(-1)^{k-1}\frac{(k-1)!}{(1+t)^k} \qquad (k=1,2,3,\ldots)$$

and therefore $$f^{(k)}(0)=(-1)^{k-1}(k-1)!.$$

Thus $$f(x)=x-\frac{x^2}{2}+\frac{x^3}{3}+\ldots+(-1)^{n-2}\frac{x^{n-1}}{n-1}+\frac{f^{(n)}(c)}{n!}x^n, \tag{1}$$

where $c$ lies between 0 and $x$. Now, if $0<x<1$ then

$$\left|\frac{f^{(n)}(c)}{n!}x^n\right|=\left|\frac{(n-1)!}{(1+c)^n n!}x^n\right|=\frac{x^n}{(1+c)^n n},$$
$$\leqslant\frac{x^n}{n}$$

because $0<c<x<1$ and $x^n/n\to 0$ as $n\to\infty$. Hence if $0<x<1$ then

$$f(x)=\log(1+x)=x-\frac{x^2}{2}+\frac{x^3}{3}\ldots=\sum_{n=1}^{\infty}(-1)^{n-1}\frac{x^n}{n}, \tag{2}$$

which is the familiar expansion of $\log(1+x)$. The same reasoning could be applied when $x=1$. In this case we observe that

$$\frac{x^n}{(1+c)^n n}=\frac{1}{(1+c)^n n}<\frac{1}{n}$$

where $0<c<x=1$. Thus the remainder tends to zero as $n\to\infty$, and the relation (2) is true for $x=1$. This is hardly surprising in view of the conclusion of Example 4 in the section following Theorem 3.3.5(a).
If, however, $-1<x<0$, then the remainder term is

$$\frac{f^{(n)}(c)}{n!}x^n=(-1)^{n-1}\frac{x^n}{(1+c)^n n},$$

and we do not know what happens to this as $n\to\infty$ because the relation $-1<x<c<0$ does not give us sufficient information about $[x/(1+c)]^n$. It can actually be proved that relation (2) is also true when $-1<x<0$, but the method we used above will not provide that proof. Moreover, if $|x|>1$, then $x^n/n\nrightarrow 0$ as $n\to\infty$ and the series on the right-hand side of (2) does not converge for any value of $x$ for which $|x|>1$.

Taylor's theorem can be used extensively to develop power series expansions for suitable functions. The reader is warned, however, to treat series expansions with care. Don't just take it for granted that the series

$$f(a)+f'(a)(x-a)+\frac{f''(a)}{2!}(x-a)^2+\ldots$$

will have the same value as $f(x)$. There are functions $f$, for which this series converges, but its sum is not $f(x)$—so be warned! The only safe way is to use Taylor's theorem correctly and check what happens to the remainder term. As a warning to the reader of what can happen we now include a further example.

3. Let $f: \mathbb{R} \to \mathbb{R}$ be defined by

$$f(x) = \begin{cases} e^{-1/x^2} & (x \neq 0), \\ 0 & (x = 0). \end{cases}$$

Then $f(x) \to f(0)$ as $x \to 0$ and $f$ is continuous at the origin. Using the rule for composite functions, we see that $f$ is continuous at all other points of $\mathbb{R}$. Now, if $x \neq 0$ then

$$\frac{f(x) - f(0)}{x - 0} = \frac{1}{x} e^{-1/x^2}$$

and
$$\frac{1}{x} e^{-1/x^2} \to 0$$

as $x \to 0$. Hence $f$ is differentiable at the origin and $f'(0) = 0$. Moreover, if $x \neq 0$, then $f'(x) = (2/x^3)e^{-1/x^2}$ using the rule for differentiating composite functions. Therefore if $x \neq 0$ then

$$\frac{f'(x) - f'(0)}{x - 0} = \frac{2}{x^4} e^{-1/x^2},$$

and this tends to zero as $x \to 0$. Hence $f''(0)$ exists and $f''(0) = 0$. Continuing in this way we can show that $f^{(k)}(0) = 0$ $(k = 3, 4, \ldots)$. Thus the series

$$f(0) + f'(0)x + \frac{f''(0)}{2!} x^2 + \ldots$$

consists entirely of zeros and its sum is zero for all real numbers $x$. Hence the series

$$f(0) + f'(0)x + \frac{f''(0)}{2!} x^2 + \ldots$$

converges for all real numbers $x$ and has sum zero, which is not equal to $f(x)$ if $x \neq 0$.

---

This final chapter has introduced the idea of differentiability and demonstrated that differential calculus has a rigorous foundation. The basic idea in the development has been the concept of a limit, a theme which has recurred repeatedly through the volume. First it appeared in the context of sequences and then later it returned in connection with functions. This notion of a limit, together with the concept of the real number system, underpins the analysis of this volume. The outcome of our work has been the development of numerous theorems—perhaps a somewhat daunting achievement for the beginner. These theorems, however, are the tools of the trade to the budding analyst. Without theorems, life would be much harder for the mathematician and problems would take much longer to solve. Alexander the Great, in his day, found that mathematics contained very many theorems, propositions and definitions and a great deal of time and effort was needed to learn them. Like his general Ptolemy, Alexander asked if there was an easier way to master mathematics, and was assured that there are no shortcuts. This is still true today. It requires time and effort to learn analysis and to learn how to apply it.

## *MISCELLANEOUS EXERCISES 5*

**1** Find the local maxima, local minima and points of inflection of $x^2(x-1)^3$.

**2** Let
$$h(x)=\frac{x^2-x-2}{(x-1)^2(x+1)^2} \qquad (x\neq \pm 1).$$
Decide whether $\lim_{x\to a}h(x)$ exists in each of the following cases: $a=-1,0,1,2$. In any case in which a limit exists find it.

**3** The function $g\colon \mathbb{R}\to\mathbb{R}$ is given by
$$g(x)=\begin{cases} x^2 & (x \text{ rational}),\\ 0 & (x \text{ irrational}).\end{cases}$$
Show that $g$ is differentiable at the origin. Show also that $g$ is discontinuous at all points of $\mathbb{R}\backslash\{0\}$.

**4** The function $g\colon \mathbb{R}\to\mathbb{R}$ is given by
$$g(x)=\begin{cases} x^2 & (x \text{ rational}),\\ x^3 & (x \text{ irrational}).\end{cases}$$
Show that $g$ is continuous at $0, 1$ and discontinuous at all other points. Where is $g$ differentiable?

**5** Decide whether the following limits exist:
$$\lim_{x\to 0}\frac{e^{x^2}-1}{\cos x-1}; \qquad \lim_{x\to 0}\left(\cot x-\frac{1}{x}\right); \qquad \lim_{x\to 0}(e^x+x)^{1/x}.$$
In any case in which the limit exists find it.

**6** Show that the equation
$$\tanh 2x+\cos(x/2)=x$$
has a solution in the interval $[-2,2]$.

**7** Let $f, g, h: [-1, +1] \to \mathbb{R}$ be given by

$$f(x) = |x|^3 \qquad (-1 \leqslant x \leqslant 1),$$
$$g(x) = x + |x| \qquad (-1 \leqslant x \leqslant 1),$$
$$h(x) = x^3 + [x] \qquad (-1 \leqslant x \leqslant 1),$$

where $[x]$ is the greatest integer which does not exceed $x$.

State whether each of the functions $f, g, h$ satisfy the conditions of the MVT. Do the conclusions of the MVT hold for any of these functions?

**8** The function $f$ is continuous on $[a, b]$ and differentiable on $(a, b)$ and $f'(x) > 0$ for all $x$ such that $a < x < b$. Show that $f$ is strictly increasing on $[a, b]$.

Hence show that for all $x > 0$,

$$x - \tfrac{1}{2}x^2 < \log(1 + x) < x - \tfrac{1}{2}x^2 + \tfrac{1}{3}x^3.$$

**9** The function $g$ is continuous on $[a, b]$ and differentiable on $(a, b)$ and $0 \leqslant g'(x) \leqslant 1$ for all $x$ such that $a < x < b$. Moreover, $g(a) = a$. Prove that if $a \leqslant x \leqslant b$, then $a \leqslant g(x) \leqslant b$. Show also that $g(b) = b$ if and only if $g(x) = x$ for all $x \in [a, b]$.

**10** Use Taylor's theorem to develop series expansion for $\sin x$, $\cos x$, $\sinh x$, $\cosh x$. Check carefully that the remainder term tends to zero as $n \to \infty$.

**11** Show that for all real numbers $\alpha$,

$$\frac{\log(1 + \alpha x)}{x} \to \alpha$$

as $x \to 0$. Deduce that

$$n \log(1 + \alpha/n) \to \alpha$$

as $n \to \infty$. Hence, show that

$$\left(1 + \frac{\alpha}{n}\right)^n \to e^\alpha$$

as $n \to \infty$.

## HINTS FOR SOLUTION OF EXERCISES

### Exercises 5.1.1

**2** For $x > 0$, $f(x) = e^x$ and so $f'(x) = e^x (x > 0)$. Similarly, $f'(x) = 1$ for $x < 0$. To prove differentiability at the origin, consider what happens to the quotient $[f(x) - f(0)]/(x - 0)$ as $x \to 0$.

**3** (e) Write $f(x) = e^{(\log x)(\log x)}$ $(x > 0)$.

## Exercises 5.2.1

**1** Use the intermediate value theorem to show there is a solution in $[0,1]$. Use Rolle's theorem to show that there cannot be more than one solution in $[0,1]$.

**2** Apply MVT to $\sin x$ on the interval $[a,b]$.

**3** Apply MVT to $\log x$ on the interval $[a,b]$.

**5** See example immediately preceding the exercises.

**6** Apply MVT to $\sin^{-1} x$ on $[a,b]$.

**7** Let $\alpha = f(0)$, then $\alpha > 0$ and $g(0) = \alpha$. Use the MVT to show that $g(3\alpha) < -\frac{1}{2}\alpha$. Then the intermediate value theorem will give the existence of $c$ for which $g(c) = 0$.)

## Miscellaneous Exercises 5

**2** Note that $x^2 - x - 2 = (x-2)(x+1)$ and therefore

$$h(x) = \frac{x-2}{(x-1)^2(x+1)} \qquad (x \neq \pm 1).$$

**3** To prove differentiability at the origin, consider what happens to $[g(x) - g(0)]/(x-0)$ as $x \to 0$. Notice that the modulus of this quotient never exceeds $|x|$ when $x \neq 0$.
Use sequences to prove the discontinuity.

**4** Remember that differentiability implies continuity, and so $g$ is not differentiable at any point where it is discontinuous.

**5** Write $\cot x - 1/x$ as $(x\cos x - \sin x)/x \sin x$ and use l'Hôpital's rule. Show that

$$\frac{\log(e^x + x)}{x} \to 2 \qquad \text{as } x \to 0$$

and deduce that $(e^x + x)^{1/x} \to e^2$ as $x \to 0$.

**6** Use the intermediate value theorem.

**7** Show that $|x|^3$ is differentiable everywhere. Show that $g$ is differentiable at all points of $[-1,1]$ except the origin. Show that $h$ is not differentiable at the origin, but is differentiable at all other points of $(-1,1)$. Since $f$ satisfies the conditions of MVT, the conclusions of MVT must hold for $f$. By definition $g(1) - g(-1) = 2$ and so $[g(1) - g(-1)]/2 = 1$, but $g'(x) = 2$ for $x > 0$ and $g'(x) = 0$ for $x < 0$ and the conclusion of the MVT does not hold. By definition $[h(1) - h(-1)]/2 = \frac{4}{2} = 2$ and $h'(x) = 3x^2$ for $0 < x < 1$. Hence $h'(\sqrt{\frac{2}{3}}) = 2$.

# ANSWERS TO EXERCISES

## Exercises 5.1.1

**3** (a) $f'(x) = 3x^2 \sinh x \sin(x^2) + x^3 \cosh x \sin(x^2) + 2x^4 \sinh x \cos(x^2)$ $(x \in \mathbb{R})$.

(b) $f'(x) = (\sin x + x \cos x)e^{x \sin x}$ $(x \in \mathbb{R})$.

(c) $f'(x) = -2x \sin[\log(1+x^2)]/(1+x^2)$ $(x \in \mathbb{R})$.

(d) $f'(x) = 2x/\sqrt{2+2x^2+x^4}$ $(x \in \mathbb{R})$.

(e) $f'(x) = 2x^{\log x - 1}\log x$ $(x > 0)$.

## Exercises 5.4.1

**1** $-2$. (Remember, you want the limit as $x \to 1$.)

**2** $32/\pi^2$. **3** $\cot 1$ (l'Hôpital's rule not appropriate!)

**4** $\sin 1$. **5** $1/\pi^2$. **6** 0. **7** $-1/e$. **8** $\frac{1}{2}$. **9** $\frac{1}{4}$. **10** $-\frac{1}{2}$.

**11** 1. **12** $-\pi$. **13** $-\frac{1}{2}$. **14** $\frac{7}{3}$. **15** 2.
**16** $\cosh(\pi/2)$ (don't use l'Hôpital).

## Miscellaneous Exercises 5

**1** Local maximum at $x = 0$; local minimum at $x = \frac{2}{5}$; point of inflection at $x = 1$.

**2** There is no limit as $x \to -1$; $h(x) \to -2$ as $x \to 0$; $h(x) \to -\infty$ as $x \to 1$; $h(x) \to 0$ as $x \to 2$.

**4** $g$ is differentiable at the origin.

**5** All the limits exist. Their values are $-2, 0, e^2$.

**7** $f$ satisfies the conditions of the MVT on $[-1, 1]$ but $g, h$ do not. The conclusions of the MVT hold for $f$ and $h$, but not for $g$.

# INDEX